PENGUIN BOOKS

THE STATE OF THE WORLD ATLAS
fifth edition

THE
STATE OF THE WORLD
ATLAS

new revised fifth edition

Michael Kidron and Ronald Segal

PENGUIN
REFERENCE

PENGUIN BOOKS

Published by the Penguin Group
Penguin Books Limited, 27 Wrights Lane,
London W8 5TZ, England
Penguin USA Inc., 375 Hudson Street,
New York, New York 10014, USA
Penguin Books Australia Limited,
Ringwood, Victoria, Australia
Penguin Books Canada Limited, 10 Alcorn Avenue,
Toronto, Ontario, Canada M4V 3B2
Penguin Books (NZ) Limited, 182–190 Wairau Road,
Auckland 10, New Zealand
Penguin Books Limited, Registered Offices:
Harmondsworth, Middlesex, England

First published 1995
10 9 8 7 6 5 4 3 2 1

Viking (hard) 0-670-86545-1
Penguin (paper) 0-14-025204-5

Produced for the Penguin Group by
Myriad Editions Limited, 32 Hanway Street,
London WIP 9DD

Edited and co-ordinated for Myriad Editions by
Anne Benewick and Candida Lacey
Graphic design by Corinne Pearlman
Text design by Pentagram Design Limited
Maps created by Angela Wilson for
Line + Line Limited, Thames Ditton, Surrey

Printed and bound in Hong Kong
Produced by Mandarin Offset Limited

CONTENTS

This fifth edition of the atlas records considerable changes. The end of the Soviet Union has led to a number of successor states, with little evidence that the lives of many citizens have been enhanced in consequence. The former satellites, those so-called people's democracies which were so conspicuously neither popular nor democratic, have gone their various ways: East Germany, for instance, to reunification with the West; Czechoslovakia to a division whose benefits appear as elusive as its purposes appeared unwholesome to outsiders; and Yugoslavia, that early breakaway satellite, to a fragmentation whose rival nationalisms have added 'ethnic cleansing' to the language of contemporary politics.

If much may seem to have changed, much has stayed the same or become simply more so. Ever more secure in its power and authority is a business culture whose boards of management pay small regard to the sovereignty of the state except to exploit it. The market is exalted as the dynamic of personal freedom and social endeavour, and the marketplace is increasingly global in its operations. Within this ascendancy, what is considered legitimate business has both a colleague and a competitor in organized crime: both recognize frontiers only in contriving how to cross them; both measure success or failure only by the bottom line. Indeed, in Russia, Ukraine and other successor states of the former USSR, such crime seems to be beating legitimate business at its own game.

Within the rich states of established parliamentary institutions, the concentration of power has proceeded apace, with sharpening economic inequalities in many of them. An increasing government impatience towards the victims of the process is revealed in state erosions of social welfare, as an increasing repressiveness looks to the building of still more prisons. What was once among the rich states a commitment, albeit largely rhetorical, to provide aid for economic development in the poor states, is now a concern for tighter immigration controls to keep the wretched of the earth in their place. Only for migrating capital is there always an asylum.

There is an exposed nerve in the system, however, and the pain that it generates grows and spreads, despite attempts to deny or belittle it. The convenience of driving a car wherever you please does not cancel the distress of watching your own children struggle with asthma. Reports of global warming become less alarmist and more alarming when freak weather patterns assail your own neighbourhood. The strictest of immigration controls are useless against the emissions of greenhouse gases half a world away. When the bosses of the insurance business become ecologists, the crisis has moved beyond the apostles of alternative lifestyles.

The approaching end of the millennium looks set to be accompanied by apocalyptic visions, which may find evidence enough in our atlas to sustain them. We remain sanguine all the same, if only because with every day that passes, the posturings and preoccupations, the divisiveness and danger, the essential disorderliness of competing states, appear more absurd. We are well aware that any call to a united human community brings the pat reply that this is unreal. Yet, surely, what is far more unreal, or should be, is that innumerable children are killed or maimed by landmines laid in some past or present war; that supposedly allied states turn upon one another over who should still further deplete the already dying sea of its fish; that humanity is busily choking on its own fumes. Our case rests.

We thank in the acknowledgements various people for the help they have so readily given us. We take pleasure in expressing here our gratitude to the editorial team at Myriad — Anne Benewick, Candida Lacey and Corinne Pearlman — for spurring us into increasing our pace when our energy flagged and for reining in our zeal when it threatened to run away with us. Above all, we thank each other, for one more episode in a collaboration that has withstood fifteen years of ageing, but whose spirit is constantly rejuvenated by a shared sense of fun as well as of purpose.

Michael Kidron
Ronald Segal

ACKNOWLEDGEMENTS

The authors would like to thank the following people for their help in connection with:

4 The Guzzlers (energy) : Fred Dixon, World Energy Council, London.

5 Infernal Combustion (vehicles): Roger Mackett, Centre for Transport Studies, University of London; John Polak, Transport Studies Unit, Oxford University; Peter White, Transport Studies Group, University of Westminster, London; Chris Batten, Andrea Lannon and Martin Rafferty, Department of Transport, London; Helen Berry, The Chartered Institute of Transport, London; and, Richard Watkins, The Highways Agency, London.

8 Diasporas: Professor Bhikhu Parekh of Hull University.

13 Connections (telecommunications): Emma Jackets and Lydia Jacobson, International Institute of Communications, London; Enzo Paci, World Tourist Office, Madrid; Dave Taylor, Office of Telecommunications (Oftel), London; and, Gregory Staple and Zachary Schrag, Telegeography Inc., Washington, D.C.

16 Missing Women (women's rights): Janet Gormek, Salim Johan and Inge Kaul at the Human Development Report office, UNDP, New York; Joann Vanek, at the UN Statistical Office, New York; Ann Hennan and Andrea Subhan, at the European Parliamentary Committee on Women's Rights office, Luxembourg; Lesley Mitchell, Association of Female Barristers, London; and, Sally Boden and Naila Kabeer, Institute of Development Studies, University of Sussex.

18 Child Exploitation (child labour): Bernie Russell, 'Free Labour World', Brussels.

19 The Killing Weed (tobacco): George Gay, World Tobacco, Redhill, Surrey; Alan Lopez, World Health Organization, Geneva; Judith Mackay, Asian Consultancy on Tobacco Control, Hong Kong; Bill Marks, U.S. Center for Disease Control, Office on Smoking and Health, Atlanta, Georgia; Richard Peto, Radcliffe Infirmary, Oxford; and, Nicholas Wald, St Bartholomew's Hospital Medical College, London.

20 Plagues New and Renewed (AIDS and tuberculosis): Dr Paul Akibumi Sato of the WHO Global Programme on AIDS and Mrs Kathleen Canon of the WHO Tuberculosis Programme.

23 The Labour Force: John Svenningsen, Director of the Bureau for Workers' Activities, and his colleague, Coen Daman, as well as other experts in the International Labour Office, Geneva.

27 Hedges and Bets (financial markets): the librarians of the Reference Library, U.S. Embassy, London.

28 The Golden Fix (narcotics): Ann Girvan, U.S. Reference Center, London.

29 Privatization: Rodney Lord, Privatisation International, London and Charles Vuylsteke, European Bank for Reconstruction and Development, London.

31 Notional Income (purchasing power): Alan Heston, Professor of Economics, University of Pennsylvania; Melinda Limjap, Division of Information, Bureau for Resources and Special Activities, United Nations, New York; Alexandra McLeod, UN Information Centre, London; and, Sarus Menon and Laura Mourino, Office of the Human Development Report, United Nations Development Program, New York.

39 Under Arms (the armed forces): Jennifer Bond, National Science Foundation, Washington, D.C.; Howard Clark, War Resisters' International, London.

40 State Terror (capital punishment and torture): Elana Dallas, Amnesty International, and Amanda Smith at the Death Penalty Information Center, Washington, D.C.

41 Smoking Guns (war and peace): Dan Smith, International Peace Research Institute (PRIO), Oslo.

42 God and Caesar (religion); Martin Palmer and Joanne O'Brien, International Consultancy on Religion, Education and Culture, Manchester.

45 Refugee Makers (refugees): Ginny Hamilton, U.S. Committee for Refugees, Washington, D.C.

47 The Three Monkeys (censorship): Philip Spender and Irena Maryniak, Index on Censorship, London.

48 Tongue-tied (language): Donald Kenrick, London, and Erik Gunnemark, Gothenburg.

50 Addition and Division (states): Alexandra McLeod, UN Information Centre, London; Carol Priestly and other staff at the Foreign and Commonwealth Office, London; and, staff at the Library, Chatham House (The Royal Institute of International Affairs), London.

Part One: INTIMATIONS OF MORTALITY

The graphic lists reported fatal radiation accidents or incidents connected with nuclear power plants and laboratories between 1945 and 1992. They range from deaths caused by the theft of radioactive material to those resulting from the major nuclear reactor accident at Chernobyl in 1986.

This tells only part of the story. In 1994, the World Health Organisation reported a substantial and continuing increase in childhood thyroid cancer in Belarus, Russia and Ukraine following the Chernobyl disaster.

Fatal accidents at military installations are not included.

Sources: International Atomic Energy Agency;
Nuclear Engineering International.

Los Alamos, U.S.A. 1945
Los Alamos, U.S.A. 1946
Vinca, Yugoslavia 1958
Los Alamos, U.S.A. 1958
Idaho Falls, U.S.A. 1961
Woods River, U.S.A. 1964
Switzerland n.d.
Mexico City, Mexico n.d.
West Germany n.d.
Rhode Island, U.S.A. n.d.
Brescia, Italy 1975
Oklahoma, U.S.A. n.d.
Algeria n.d.
Norway n.d.
Constituentes, Argentina 1983
Morocco n.d.
Chernobyl, Ukraine 1986
Goiania, Brazil n.d.
Xin Zhou, China 1992

GREENLAND
(Den)

ICELAND

CANADA

UNITED STATES
OF AMERICA

BERMUDA

MEXICO

36%

17%
BAHAMAS

CUBA

DOMINICAN
REPUBLIC

BELIZE   JAMAICA   HAITI
GUATEMALA   HONDURAS
EL SALVADOR
NICARAGUA   23%
COSTA RICA
PANAMA

17%
BARBADOS
TRINIDAD & TOBAGO

VENEZUELA   GUYANA
COLOMBIA   SURINAME
            FRENCH GUIANA (Fr)

ECUADOR

PACIFIC
OCEAN

ATLANTIC
OCEAN

CAPE VERDE

NORWAY   SWEDEN

10%
UNITED   DENMARK
KINGDOM
IRELAND
NETH.   POLAND
BEL   GERMANY
                CZECH   SLO
FRANCE   AUS   HUNG
        ITALY   B-H/YUG
PORTUGAL   SPAIN   ALB
                              GREE
MALTA 46%
TUNISIA
MOROCCO   13%
                        11%   16%
WESTERN SAHARA MAURITANIA   ALGERIA   LIBYA
                    10%
SENEGAL   23%   12%   11%   13%
GAMBIA   MALI BURKINA NIGER CHAD
GUINEA-BISSAU   FASO
    GUINEA   BENIN NIGERIA
SIERRA LEONE CÔTE d'
LIBERIA   IVOIRE GHANA   CAR
              TOGO
           EQUATORIAL GUINEA   CAMEROON
                    GABON
                   CONGO   ZAI

ANGOLA

NAMIBIA

10%
SOUTH
AFRICA

PERU

BOLIVIA

CHILE   PARAGUAY

URUGUAY

ARGENTINA

BRAZIL

10%

THREATENED SPECIES
1990s

10% or over of known mammal
species are threatened

Highest: Madagascar, 48%; Cuba, 36%; Fiji, 25%;
Jamaica, Mauritania, 23%

10% or over of known bird species
are threatened

Highest: Malta, 46%; Kuwait, 26%; Saudi Arabia, 20%

Source: World Resources Institute, *World Resources 1994-95.*

**PROPORTION OF LAND AREA SUBJECT TO
HIGH HUMAN DISTURBANCE**
from permanent agricultural or
urban development, soil degradation,
and other factors
1993 percentages

- 80%
- 60%
- 40%
- 20%
- no data

Highest: Denmark, Ireland, 100%; Netherlands, 98%; Burundi, 97%

THREATENED BIRDS *1990s*
percentages of total number of species known
Source: World Resources Institute, *World Resources 1994-95.*

3 countries — 15%
8 countries — 10%
44 countries — 5%
70 countries

"...The sedge is wither'd from the lake,
And no birds sing."
- John Keats

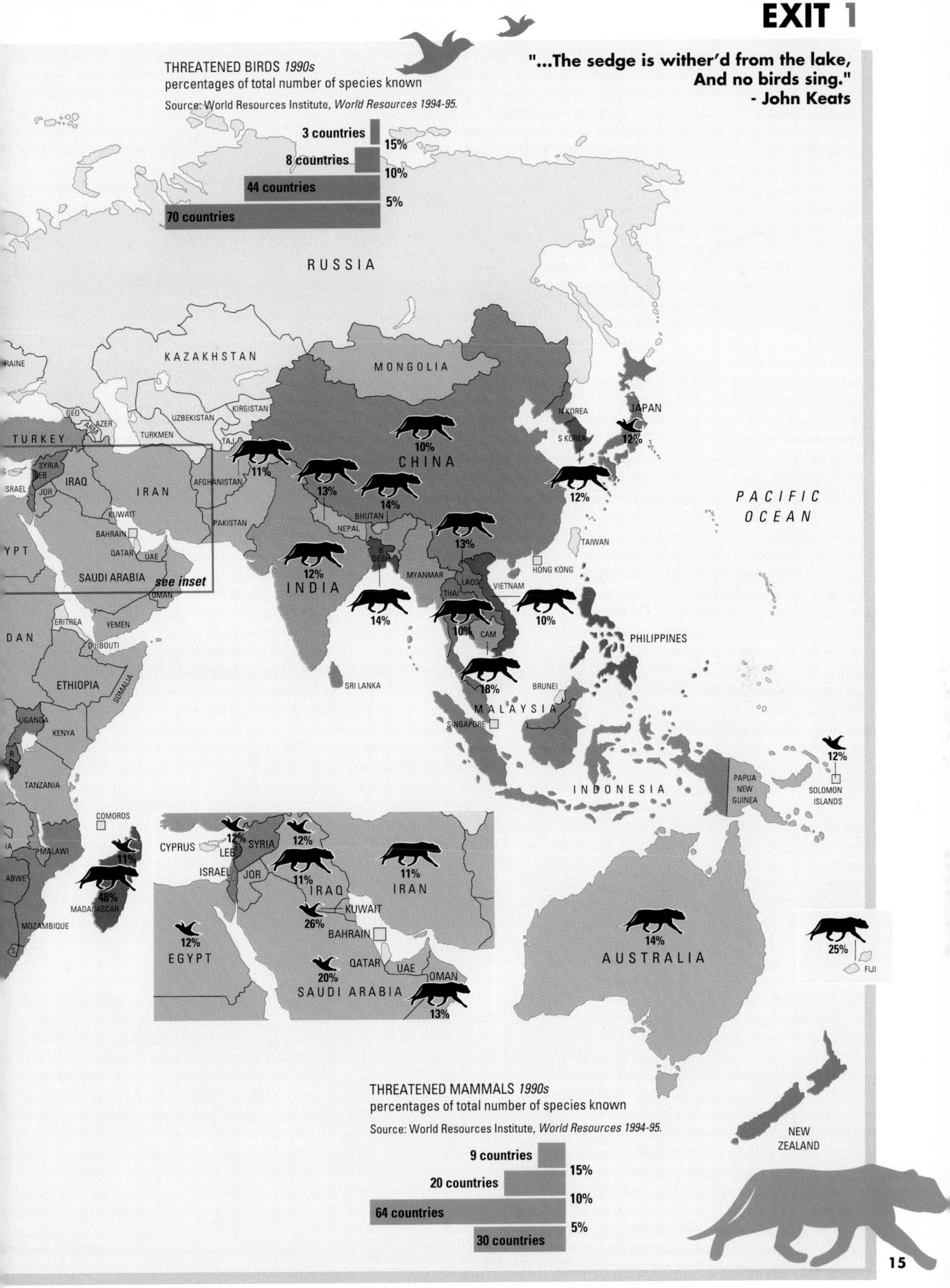

RUSSIA

KAZAKHSTAN

MONGOLIA

N.KOREA
S.KOREA

JAPAN 12%

UKRAINE
GEO AZER
TURKEY
UZBEKISTAN KIRGISTAN
TURKMEN TAJ.
SYRIA
DEB
ISRAEL JOR IRAQ
IRAN
AFGHANISTAN
PAKISTAN

11%
10%
CHINA
13%
14%
13%
12%

KUWAIT
BAHRAIN
QATAR UAE
SAUDI ARABIA *see inset*
OMAN

BHUTAN
NEPAL
B.DESH
INDIA
MYANMAR
THAI LAOS
VIETNAM
CAM

12%
14%
10%
10%
18%

TAIWAN
HONG KONG
PHILIPPINES

PACIFIC
OCEAN

EGYPT
ERITREA YEMEN
DJIBOUTI
DAN
ETHIOPIA SOMALIA
UGANDA
KENYA
TANZANIA

SRI LANKA
BRUNEI
MALAYSIA
SINGAPORE

INDONESIA
PAPUA NEW GUINEA

SOLOMON ISLANDS 12%

COMOROS
MALAWI 11%
ABWE 48%
MADAGASCAR
MOZAMBIQUE

CYPRUS 12% SYRIA 12%
LEB
ISRAEL 11%
JOR IRAQ
KUWAIT 26%
BAHRAIN
QATAR UAE
SAUDI ARABIA 20%
OMAN 13%

IRAN 11%

EGYPT 12%

AUSTRALIA 14%

FIJI 25%

THREATENED MAMMALS *1990s*
percentages of total number of species known
Source: World Resources Institute, *World Resources 1994-95.*

9 countries — 15%
20 countries — 10%
64 countries — 5%
30 countries

NEW ZEALAND

15

Michael Kidron and Ronald Segal *The State of the World Atlas* 5th edition

Northwest Territories

C A N A D A

UNITED STATES
OF AMERICA

California

**Mississippi
Basin**

MEXICO

CUBA

BAHAMAS

JAMAICA

DOMINICAN
REPUBLIC

HAITI

PUERTO RICO (U.S.)

BELIZE

GUATEMALA

EL SALVADOR

HONDURAS

NICARAGUA

COSTA RICA

PANAMA

TRINIDAD & TOBAGO

VENEZUELA

GUYANA

SURINAME

FRENCH GUIANA (Fr)

COLOMBIA

ECUADOR

PERU

BRAZIL

BOLIVIA

PARAGUAY

CHILE

URUGUAY

ARGENTINA

ATLANTIC
OCEAN

NORWAY

SWEDEN

FINLAND

IRELAND

UNITED
KINGDOM

DENMARK

ESTONIA

LATVIA

LITHUANIA

BELARUS

NETH

BEL

POLAND

GERMANY

CZECH
REPUBLIC

SLOVAK

UKRAINE

FRANCE

SWITZ

AUSTRIA

HUNGARY

SLO

CROATIA

ROMANIA

B - H

YUG

ITALY

ALB

M

BULGARIA

PORTUGAL

SPAIN

GREECE

MOROCCO

TUNISIA

ALGERIA

LIBYA

WESTERN SAHARA

MAURITANIA

MALI

NIGER

CHAD

CAPE VERDE

SENEGAL

GAMBIA

GUINEA-BISSAU

GUINEA

BURKINA
FASO

SIERRA LEONE

CÔTE d'
IVOIRE

GHANA

BENIN

NIGERIA

LIBERIA

TOGO

CAMEROON

CAR

EQUATORIAL GUINEA

GABON

ZAI

CONGO

ANGOLA

NAMIBIA

SOU
AFR

HAWAII

DISAPPEARING FORESTS
1981-90 annual percentage rates

2% decline in forests

1%

0.5%

marginal or no change

measurable reforestation

no data

Highest deforestation: Jamaica, 5.3%;
Haiti, 3.9%; Bangladesh, 3.3%

ENVIRONMENTAL DISASTERS CAUSED
BY FREAK WEATHER PATTERNS *1987-94*

major fires

major storms

drought

floods

coral bleaching due to rise in sea temperature

Sources: Greenpeace, *Climate Time Bomb*, 1994; World
Resources Institute, *World Resources 1994-95*; press reports.

STATES REFORESTING AT ANNUAL RATE OF
0.5 PERCENT OR ABOVE *1981-90*

0.5%

0.6%

1.1%

1.3%

Germany
Hungary
Portugal

Switzerland

UK

Ireland

Source: World Resources Institute, *World Resources 1994-95*.

16

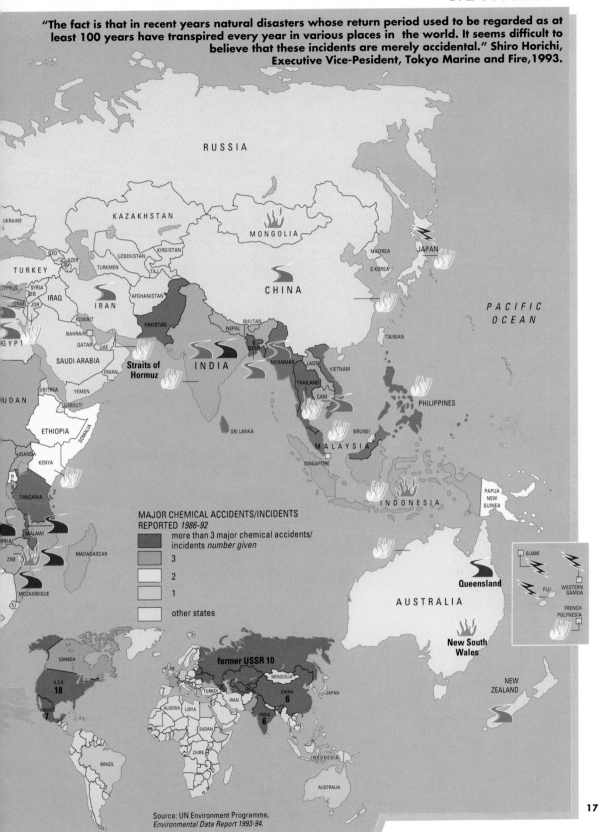

"The fact is that in recent years natural disasters whose return period used to be regarded as at least 100 years have transpired every year in various places in the world. It seems difficult to believe that these incidents are merely accidental." Shiro Horichi, Executive Vice-Pesident, Tokyo Marine and Fire,1993.

RUSSIA

KAZAKHSTAN

MONGOLIA

CHINA

JAPAN

PACIFIC OCEAN

TURKEY

IRAN

PAKISTAN

INDIA

Straits of Hormuz

SAUDI ARABIA

ETHIOPIA

PHILIPPINES

MALAYSIA

SINGAPORE

INDONESIA

PAPUA NEW GUINEA

MAJOR CHEMICAL ACCIDENTS/INCIDENTS REPORTED *1986-92*

more than 3 major chemical accidents/ incidents *number given*

3

2

1

other states

Queensland

AUSTRALIA

New South Wales

GUAM

FIJI

WESTERN SAMOA

FRENCH POLYNESIA

NEW ZEALAND

CANADA

former USSR 10

U.S.A 18

MEXICO 7

CHINA 6

INDIA 6

BRAZIL

Source: UN Environment Programme, Environmental Data Report 1993-94.

CANADA

UNITED STATES
OF AMERICA
19.1%

UNITED
KINGDOM

DENMARK

EST
LAT
LITH

BELARUS

POLAND

UKRAINE

NETH

GERMANY

HUNGARY

FRANCE

ROMANIA

MOL

BEL

BULG

former YUG

GEO

TURKME

SPAIN

ITALY

GREECE

TURKEY

AZER

PAKISTAN

AUSTRIA

IRAN

IRAQ

MEXICO

ALGERIA

EGYPT

SAUDI
ARABIA

NIGERIA

COLOMBIA

VENEZUELA

ZAIRE

ECUADOR

PERU

SOUTH
AFRICA

BOLIVIA

BRAZIL

ARGENTINA

average temperature
at beginning of 20th century

-1°C

1 million years ago

-2°C

-3°C

-4°C

10,000 years ago

previous ice ages

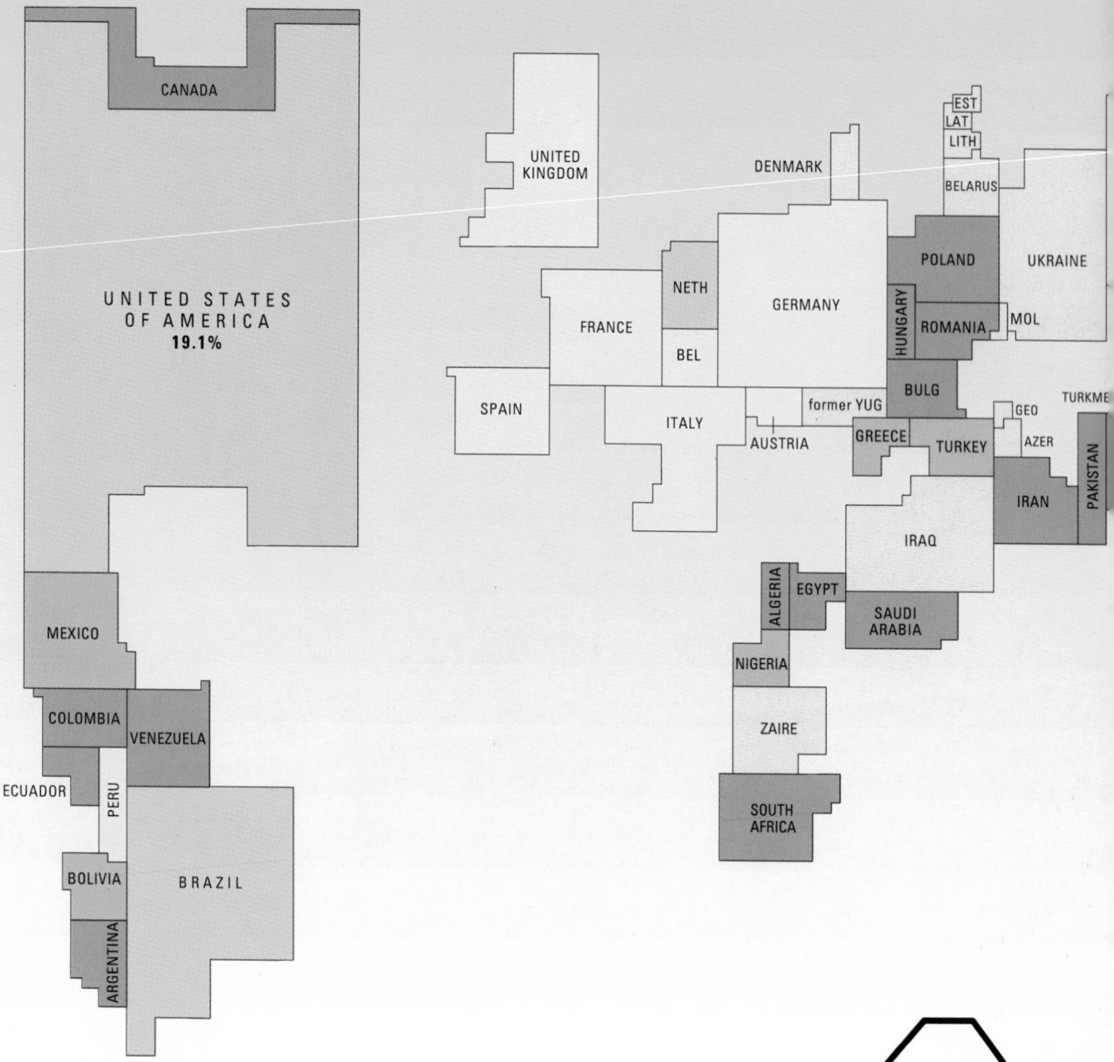

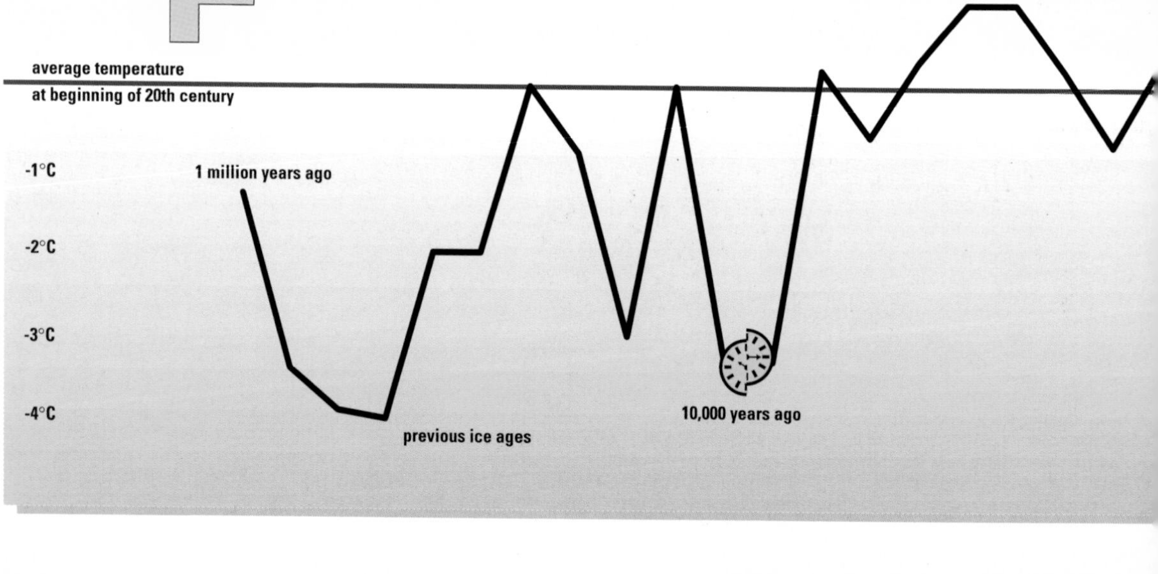

World temperatures are forecast to rise dramatically by 2100. Some islands will sink under the ocean and vast areas of land will be permanently inundated. Deserts and tropical diseases will spread.

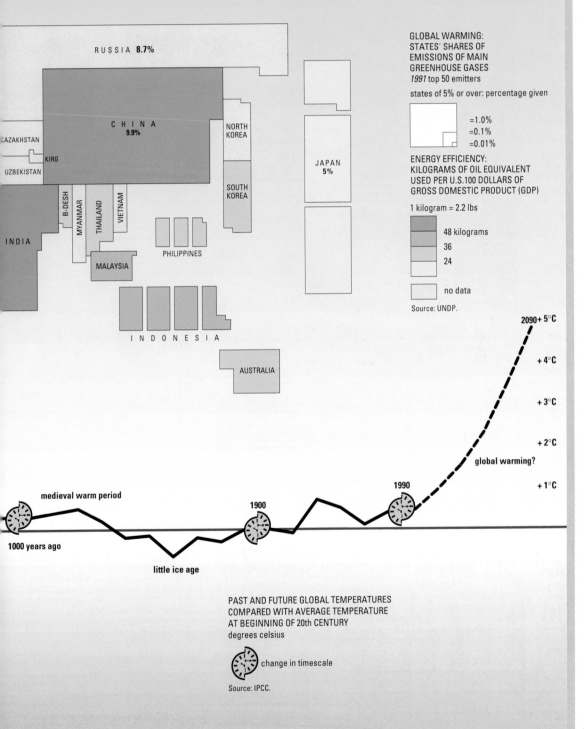

RUSSIA **8.7%**

CHINA **9.9%**

KAZAKHSTAN

KIRG

UZBEKISTAN

NORTH KOREA

SOUTH KOREA

JAPAN **5%**

B-DESH

MYANMAR

THAILAND

VIETNAM

INDIA

PHILIPPINES

MALAYSIA

I N D O N E S I A

AUSTRALIA

GLOBAL WARMING:
STATES' SHARES OF
EMISSIONS OF MAIN
GREENHOUSE GASES
1991 top 50 emitters

states of 5% or over: percentage given

=1.0%
=0.1%
=0.01%

ENERGY EFFICIENCY:
KILOGRAMS OF OIL EQUIVALENT
USED PER U.S.100 DOLLARS OF
GROSS DOMESTIC PRODUCT (GDP)

1 kilogram = 2.2 lbs

48 kilograms
36
24

no data

Source: UNDP.

2090+ 5°C

+ 4°C

+ 3°C

+ 2°C

global warming?

1990

+ 1°C

medieval warm period

1900

1000 years ago

little ice age

PAST AND FUTURE GLOBAL TEMPERATURES
COMPARED WITH AVERAGE TEMPERATURE
AT BEGINNING OF 20th CENTURY
degrees celsius

change in timescale

Source: IPCC.

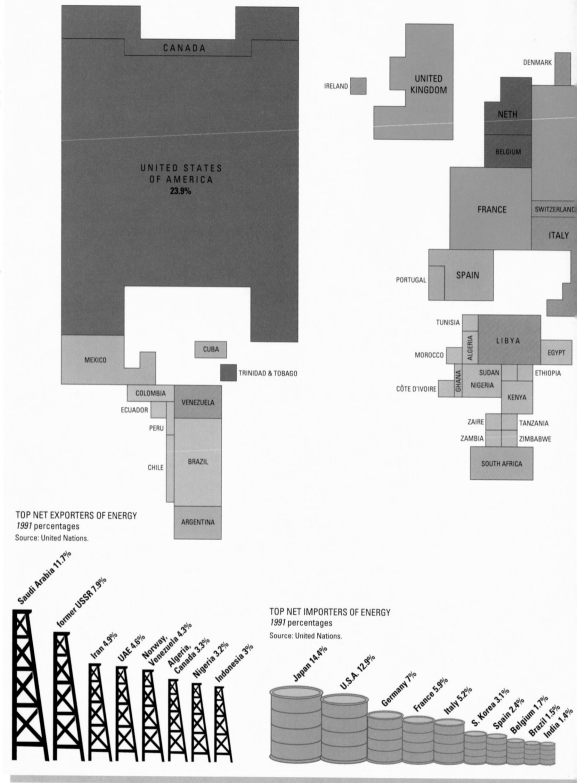

CANADA

UNITED STATES
OF AMERICA
23.9%

IRELAND

UNITED
KINGDOM

DENMARK

NETH

BELGIUM

FRANCE

SWITZERLAND

ITALY

PORTUGAL

SPAIN

MEXICO

CUBA

TRINIDAD & TOBAGO

COLOMBIA

ECUADOR

VENEZUELA

PERU

CHILE

BRAZIL

TUNISIA

ALGERIA

LIBYA

EGYPT

MOROCCO

CÔTE D'IVOIRE

GHANA

SUDAN

NIGERIA

ETHIOPIA

KENYA

ZAIRE

TANZANIA

ZAMBIA

ZIMBABWE

SOUTH AFRICA

TOP NET EXPORTERS OF ENERGY
1991 percentages
Source: United Nations.

Saudi Arabia 11.7%

former USSR 7.9%

Iran 4.9%

UAE 4.6%

Norway,
Venezuela 4.3%

Algeria,
Canada 3.3%

Nigeria 3.2%

Indonesia 3%

ARGENTINA

TOP NET IMPORTERS OF ENERGY
1991 percentages
Source: United Nations.

Japan 14.4%

U.S.A. 12.9%

Germany 7%

France 5.9%

Italy 5.2%

S. Korea 3.1%

Spain 2.4%

Belgium 1.7%

Brazil 1.5%

India 1.4%

The average person burns two and half times more energy in the U.S.A. than in Japan or Switzerland.

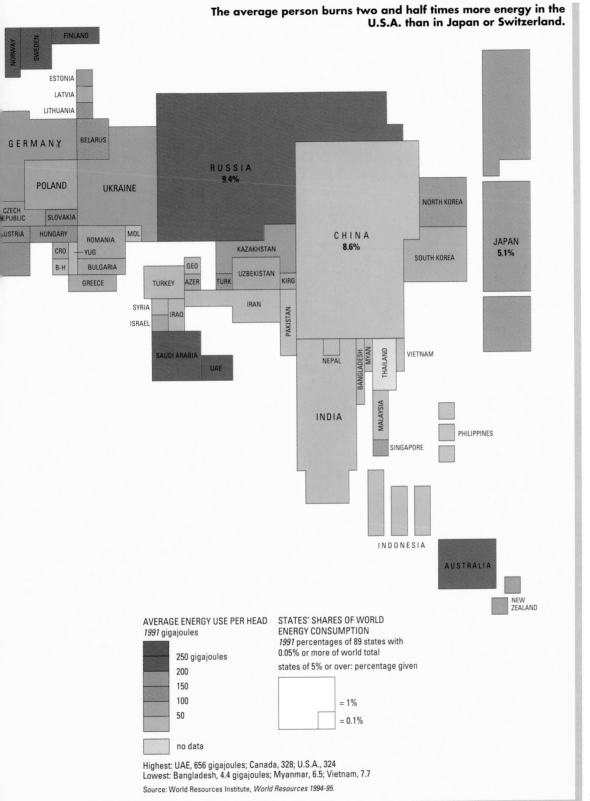

NORWAY
SWEDEN
FINLAND
ESTONIA
LATVIA
LITHUANIA
GERMANY
BELARUS
POLAND
UKRAINE
RUSSIA
9.4%
NORTH KOREA
CZECH REPUBLIC
SLOVAKIA
AUSTRIA
HUNGARY
MOL
CHINA
8.6%
JAPAN
5.1%
ROMANIA
KAZAKHSTAN
SOUTH KOREA
CRO
YUG
B-H
BULGARIA
GEO
UZBEKISTAN
GREECE
TURKEY
AZER
TURK.
KIRG
SYRIA
IRAN
ISRAEL
IRAQ
PAKISTAN
SAUDI ARABIA
UAE
NEPAL
BANGLADESH
MYAN
THAILAND
VIETNAM
INDIA
MALAYSIA
PHILIPPINES
SINGAPORE
INDONESIA
AUSTRALIA
NEW ZEALAND

AVERAGE ENERGY USE PER HEAD
1991 gigajoules

- 250 gigajoules
- 200
- 150
- 100
- 50

no data

STATES' SHARES OF WORLD ENERGY CONSUMPTION
1991 percentages of 89 states with 0.05% or more of world total

states of 5% or over: percentage given

= 1%
= 0.1%

Highest: UAE, 656 gigajoules; Canada, 328; U.S.A., 324
Lowest: Bangladesh, 4.4 gigajoules; Myanmar, 6.5; Vietnam, 7.7

Source: World Resources Institute, *World Resources 1994-95.*

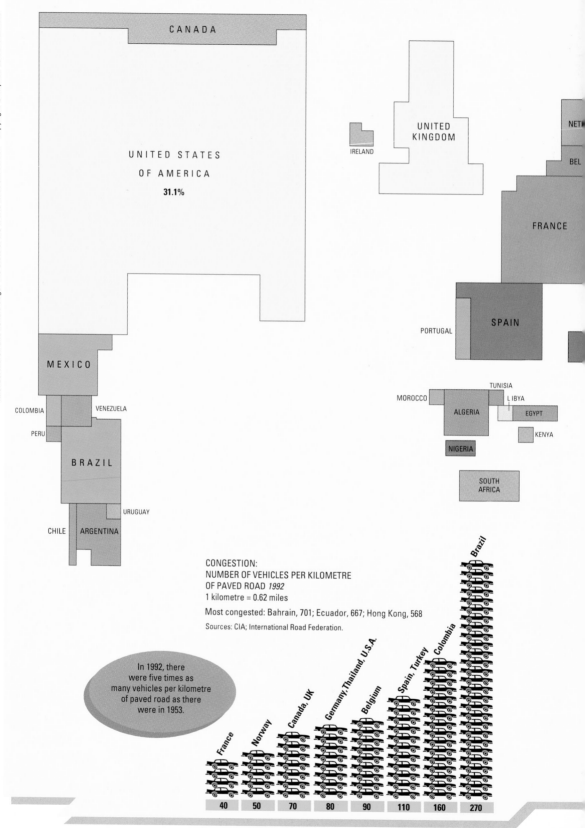

CANADA

UNITED STATES

OF AMERICA

31.1%

MEXICO

COLOMBIA

VENEZUELA

PERU

BRAZIL

URUGUAY

CHILE   ARGENTINA

IRELAND

UNITED
KINGDOM

NET

BEL

FRANCE

PORTUGAL   SPAIN

TUNISIA

MOROCCO   LIBYA

ALGERIA   EGYPT

                KENYA

NIGERIA

SOUTH
AFRICA

CONGESTION:
NUMBER OF VEHICLES PER KILOMETRE
OF PAVED ROAD *1992*
1 kilometre = 0.62 miles

Most congested: Bahrain, 701; Ecuador, 667; Hong Kong, 568

Sources: CIA; International Road Federation.

In 1992, there
were five times as
many vehicles per kilometre
of paved road as there
were in 1953.

Brazil

Germany, Thailand, U.S.A.

Spain, Turkey

Colombia

Canada, UK

Belgium

France

Norway

| 40 | 50 | 70 | 80 | 90 | 110 | 160 | 270 |

In 1950 there was one car for every 46 people in the world. In 1970 there was one for every 18; by 1994, one for every 12. Cars are running us down.

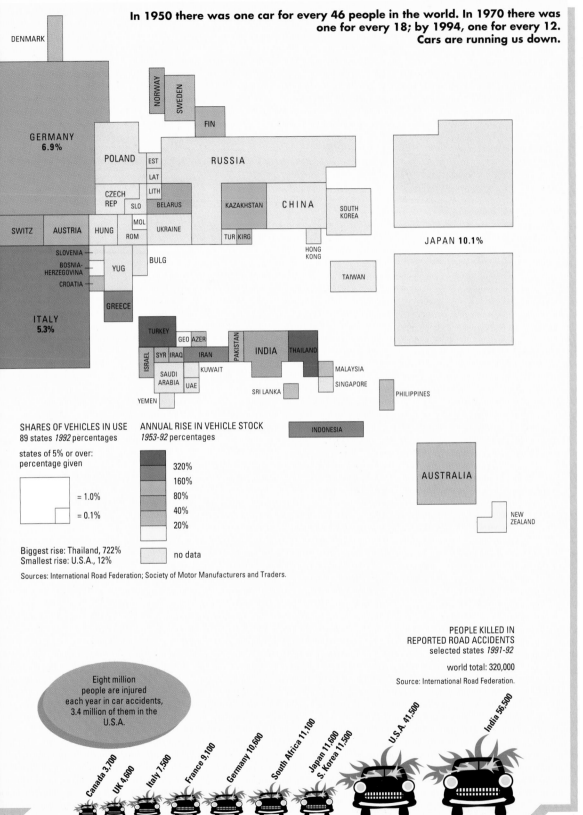

DENMARK

NORWAY

SWEDEN

FIN

GERMANY
6.9%

POLAND

EST

LAT

LITH

RUSSIA

CZECH
REP

SLO

BELARUS

KAZAKHSTAN

CHINA

SOUTH
KOREA

SWITZ

AUSTRIA

HUNG

MOL

UKRAINE

ROM

TUR KIRG

JAPAN 10.1%

SLOVENIA

BOSNIA-
HERZEGOVINA

CROATIA

YUG

BULG

HONG
KONG

TAIWAN

ITALY
5.3%

GREECE

TURKEY

GEO AZER

PAKISTAN

INDIA

THAILAND

ISRAEL

SYR IRAQ

IRAN

MALAYSIA

SAUDI
ARABIA

UAE

KUWAIT

SINGAPORE

SRI LANKA

PHILIPPINES

YEMEN

INDONESIA

AUSTRALIA

NEW
ZEALAND

SHARES OF VEHICLES IN USE
89 states *1992* percentages

states of 5% or over:
percentage given

☐ = 1.0%
☐ = 0.1%

Biggest rise: Thailand, 722%
Smallest rise: U.S.A., 12%

Sources: International Road Federation; Society of Motor Manufacturers and Traders.

ANNUAL RISE IN VEHICLE STOCK
1953-92 percentages

320%
160%
80%
40%
20%

no data

**PEOPLE KILLED IN
REPORTED ROAD ACCIDENTS**
selected states *1991-92*

world total: 320,000

Source: International Road Federation.

Eight million
people are injured
each year in car accidents,
3.4 million of them in the
U.S.A.

Canada 3,700

UK 4,600

Italy 7,500

France 9,100

Germany 10,600

South Africa 11,100

Japan 11,600

S. Korea 11,500

U.S.A. 41,500

India 56,500

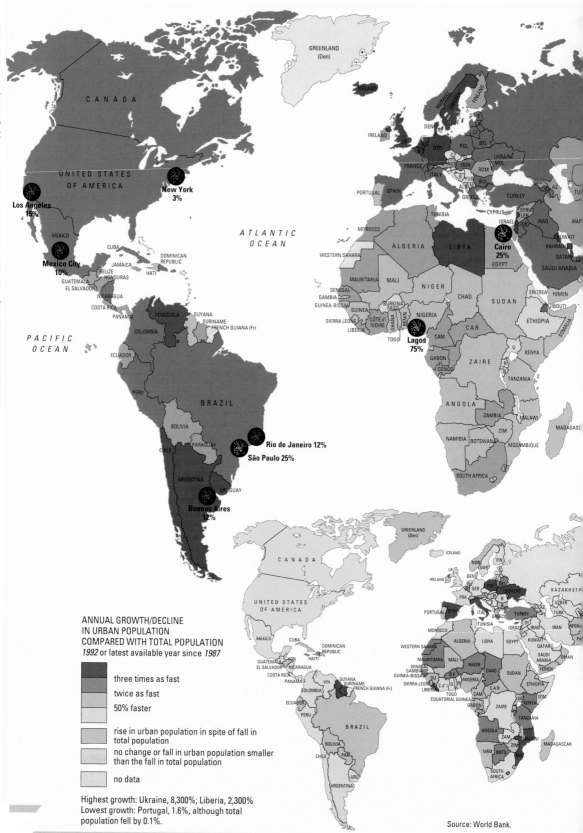

Michael Kidron and Ronald Segal *The State of the World Atlas* 5th edition   Copyright © Myriad Editions Limited

GREENLAND
(Den)

ICELAND

C A N A D A

UNITED STATES
OF AMERICA

**Los Angeles
15%**

**New York
3%**

MEXICO

**Mexico City
10%**

CUBA
JAMAICA
BELIZE
HONDURAS
GUATEMALA
EL SALVADOR
NICARAGUA
COSTA RICA
PANAMA

DOMINICAN
REPUBLIC
HAITI

ATLANTIC
OCEAN

PACIFIC
OCEAN

VENEZUELA
COLOMBIA
GUYANA
SURINAME
FRENCH GUIANA (Fr)

ECUADOR

PERU

B R A Z I L

BOLIVIA

PARAGUAY

CHILE

ARGENTINA

URUGUAY

Rio de Janeiro 12%

São Paulo 25%

**Buenos Aires
12%**

NORWAY
SWEDEN
FINLAND
DEN
IRELAND
UK
GER
BEL
POL
FRANCE
AUS HUN
ITALY
B YUG
ALB BUL
PORTUGAL SPAIN
GREECE
TURKEY
CYPRUS
SYRIA
LEB
ISRAEL JOR
IRAQ
IRAN
AZ
GEO

LI
CZ
UKRAINE
MOL
ROM

KUWAIT
BAHRAIN
QATAR
SAUDI ARABIA

TUNISIA
MOROCCO
WESTERN SAHARA
ALGERIA
LIBYA
**Cairo
25%**
EGYPT

MAURITANIA
MALI
NIGER
CHAD
SUDAN
ERITREA
YEMEN
DJIBOUTI

SENEGAL
GAMBIA
GUINEA-BISSAU
GUINEA
SIERRA LEONE
CÔTE d'
IVOIRE
LIBERIA
BURKINA
FASO
GHANA
TOGO
BENIN
NIGERIA
**Lagos
75%**
CAM
CAR
ETHIOPIA
SOMALIA

GABON
CONGO
ZAIRE
U
R
KENYA
TANZANIA

ANGOLA
ZAMBIA
ZIM
MALAWI
MADAGASC

NAMIBIA
BOTSWANA
MOZAMBIQUE

SOUTH AFRICA

**ANNUAL GROWTH/DECLINE
IN URBAN POPULATION
COMPARED WITH TOTAL POPULATION**
1992 or latest available year since *1987*

three times as fast

twice as fast

50% faster

rise in urban population in spite of fall in
total population

no change or fall in urban population smaller
than the fall in total population

no data

GREENLAND
(Den)
ICELAND

C A N A D A

NOR SWE
FIN
UK
DEN
IRELAND
GER POL
FRA
B
R
GREECE
KAZAKHSTAN
UZBEK
TURK

UNITED STATES
OF AMERICA

PORTUGAL SPAIN
ITALY
TURKEY
ISRAEL IRAQ
IRAN
AFGH.

MEXICO
CUBA
DOMINICAN
REPUBLIC
HAITI
MOROCCO
WESTERN SAHARA
ALGERIA
LIBYA
EGYPT
KUWAIT
QATAR
SAUDI
ARABIA
YEMEN
OMAN

GUATEMALA
EL SALVADOR
NICARAGUA
COSTA RICA
PANAMA

VEN
GUYANA
SURINAME
FRENCH GUIANA (Fr)
COLOMBIA
ECUADOR
PERU

MAURITANIA
MALI
NIGER
CHAD
SUDAN
ETHIOPIA

SENEGAL
GAMBIA
GUINEA-BISSAU
SIERRA LEONE
C d'
G
TOGO
EQUATORIAL GUINEA
NIGERIA
CAR
CAM
GABON
ZAIRE
UG
KENYA
TANZANIA

B R A Z I L

BOLIVIA
PAR
CHILE
URU
ARGENTINA

ANGOLA
ZAM
NAM BOTS
ZIM
MALAWI
MADAGASCAR

SOUTH
AFRICA

Highest growth: Ukraine, 8,300%; Liberia, 2,300%.
Lowest growth: Portugal, 1.6%, although total
population fell by 0.1%.

Source: World Bank.

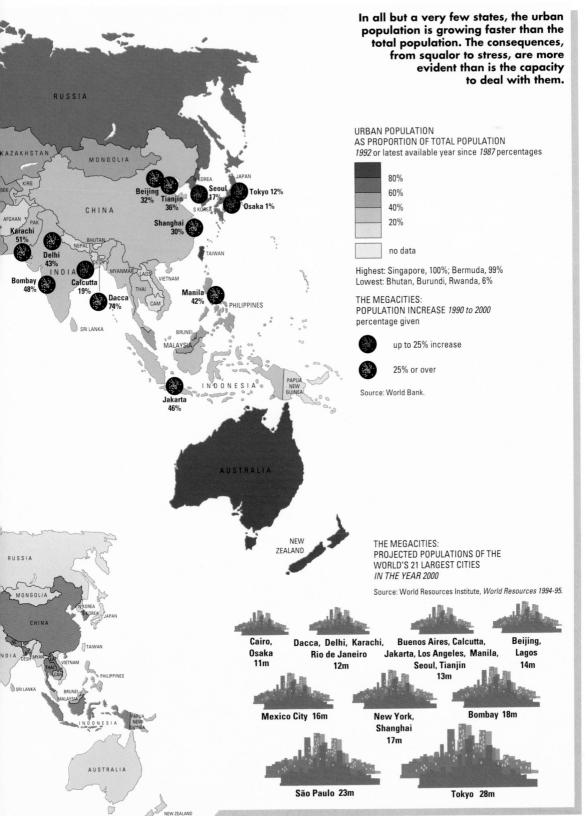

In all but a very few states, the urban population is growing faster than the total population. The consequences, from squalor to stress, are more evident than is the capacity to deal with them.

URBAN POPULATION
AS PROPORTION OF TOTAL POPULATION
1992 or latest available year since *1987* percentages

- 80%
- 60%
- 40%
- 20%

no data

Highest: Singapore, 100%; Bermuda, 99%
Lowest: Bhutan, Burundi, Rwanda, 6%

THE MEGACITIES:
POPULATION INCREASE *1990 to 2000*
percentage given

up to 25% increase

25% or over

Source: World Bank.

THE MEGACITIES:
PROJECTED POPULATIONS OF THE
WORLD'S 21 LARGEST CITIES
IN THE YEAR 2000

Source: World Resources Institute, *World Resources 1994-95*.

Cairo, Osaka 11m

Dacca, Delhi, Karachi, Rio de Janeiro 12m

Buenos Aires, Calcutta, Jakarta, Los Angeles, Manila, Seoul, Tianjin 13m

Beijing, Lagos 14m

Mexico City 16m

New York, Shanghai 17m

Bombay 18m

São Paulo 23m

Tokyo 28m

Map labels:
RUSSIA
KAZAKHSTAN
MONGOLIA
KIRG
UZBEK
AFGHAN PAK
CHINA
N KOREA
JAPAN
Beijing 32%
Tianjin 36%
Seoul 17%
Tokyo 12%
Osaka 1%
S KOREA
Shanghai 30%
Karachi 51%
Delhi 43%
BHUTAN
NEPAL
DESH
INDIA
MYANMAR LAOS
VIETNAM
TAIWAN
Bombay 48%
Calcutta 19%
THAI
CAM
Dacca 74%
Manila 42%
PHILIPPINES
SRI LANKA
BRUNEI
MALAYSIA
INDONESIA
PAPUA NEW GUINEA
Jakarta 46%
AUSTRALIA
NEW ZEALAND

People are not equal citizens within the world of states.
The average billionaire collects U.S. $200 million a year,
or as much as two million average Indians or Haitians or
six million average Ethiopians or Zaireans.

The top 101 private fortunes in the world, owned by 223
billionaire families, together claim income and capital
gains of U.S.$45.5 billion per year, the equivalent of 2.6
percent of world income.

Sources: *Fortune;* UN Development Programme.

223 billionaire families claim 2.6% of world income per year

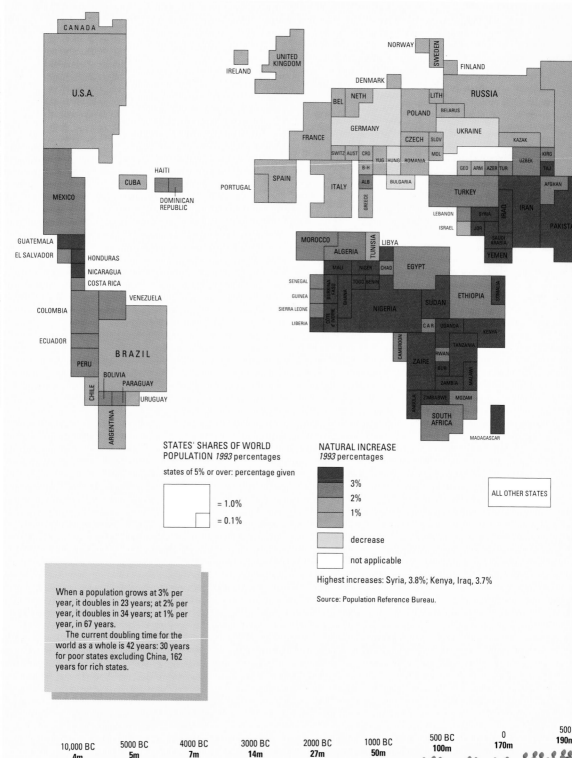

STATES' SHARES OF WORLD
POPULATION *1993* percentages

states of 5% or over: percentage given

☐ = 1.0%

▫ = 0.1%

NATURAL INCREASE
1993 percentages

- 3%
- 2%
- 1%
- decrease
- not applicable

ALL OTHER STATES

Highest increases: Syria, 3.8%; Kenya, Iraq, 3.7%

Source: Population Reference Bureau.

When a population grows at 3% per
year, it doubles in 23 years; at 2% per
year, it doubles in 34 years; at 1% per
year, in 67 years.
   The current doubling time for the
world as a whole is 42 years: 30 years
for poor states excluding China, 162
years for rich states.

10,000 BC	5000 BC	4000 BC	3000 BC	2000 BC	1000 BC	500 BC	0	500
4m	5m	7m	14m	27m	50m	100m	170m	190m

There are 5,500 million people on earth and many more on the way.

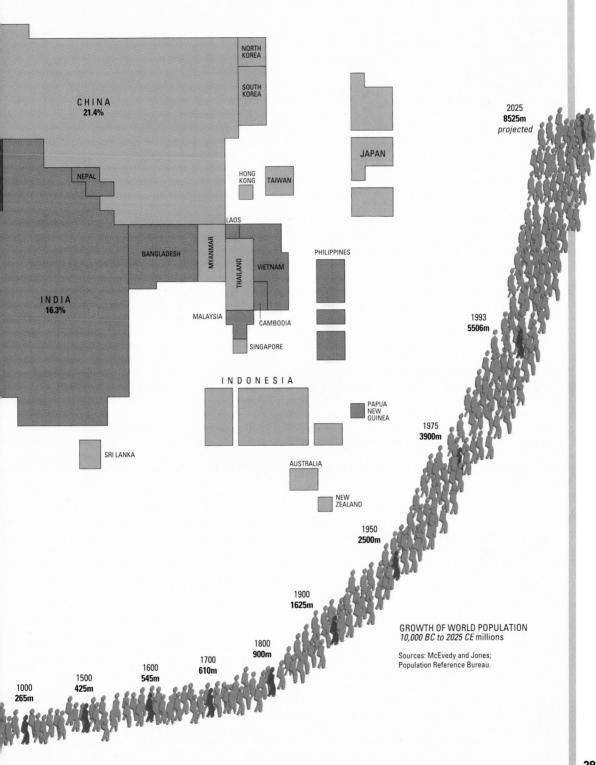

NORTH KOREA

SOUTH KOREA

CHINA
21.4%

JAPAN

NEPAL

HONG KONG

TAIWAN

LAOS

BANGLADESH

MYANMAR

THAILAND

VIETNAM

PHILIPPINES

INDIA
16.3%

MALAYSIA

CAMBODIA

SINGAPORE

INDONESIA

PAPUA NEW GUINEA

SRI LANKA

AUSTRALIA

NEW ZEALAND

2025
8525m
projected

1993
5506m

1975
3900m

1950
2500m

1900
1625m

1800
900m

1700
610m

1600
545m

1500
425m

1000
265m

GROWTH OF WORLD POPULATION
10,000 BC to 2025 CE millions

Sources: McEvedy and Jones;
Population Reference Bureau.

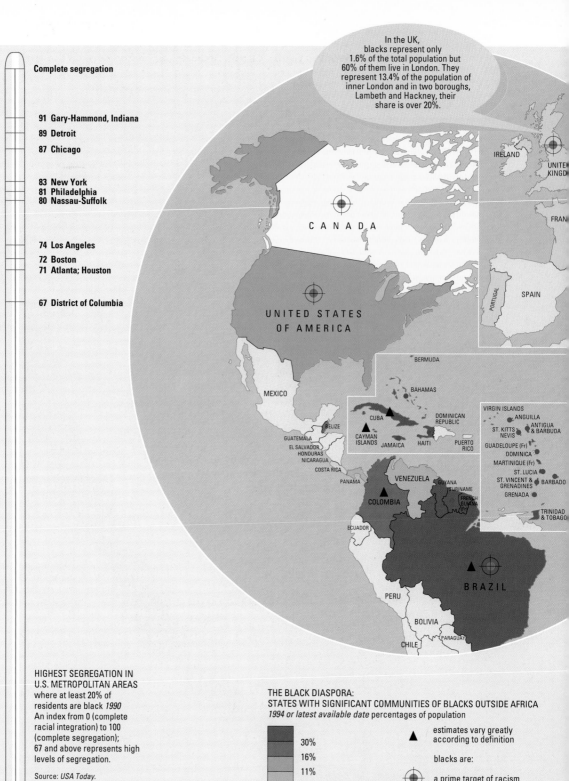

Complete segregation

91 Gary-Hammond, Indiana

89 Detroit

87 Chicago

83 New York
81 Philadelphia
80 Nassau-Suffolk

74 Los Angeles

72 Boston

71 Atlanta; Houston

67 District of Columbia

In the UK,
blacks represent only
1.6% of the total population but
60% of them live in London. They
represent 13.4% of the population of
inner London and in two boroughs,
Lambeth and Hackney, their
share is over 20%.

**HIGHEST SEGREGATION IN
U.S. METROPOLITAN AREAS**
where at least 20% of
residents are black *1990*
An index from 0 (complete
racial integration) to 100
(complete segregation);
67 and above represents high
levels of segregation.

Source: *USA Today.*

Complete integration

**THE BLACK DIASPORA:
STATES WITH SIGNIFICANT COMMUNITIES OF BLACKS OUTSIDE AFRICA**
1994 or latest available date percentages of population

30%

16%

11%

6%

1%

other states or no data

▲ estimates vary greatly
according to definition

blacks are:

⊕ a prime target of racism

● electorally dominant

Sources: Kurian; Minority Rights Group; Segal;
UK Census of Population; regional and national
handbooks and yearbooks.

Some diasporas prosper. Some do not. Either way, they are often the target of resentment and even violence.

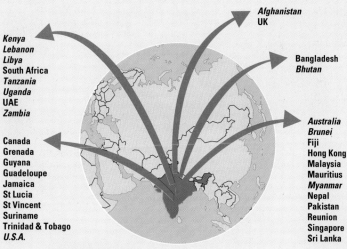

Afghanistan
UK

Kenya
Lebanon
Libya
South Africa
Tanzania
Uganda
UAE
Zambia

Bangladesh
Bhutan

Canada
Grenada
Guyana
Guadeloupe
Jamaica
St Lucia
St Vincent
Suriname
Trinidad & Tobago
U.S.A.

Australia
Brunei
Fiji
Hong Kong
Malaysia
Mauritius
Myanmar
Nepal
Pakistan
Reunion
Singapore
Sri Lanka

THE INDIAN DIASPORA:
STATES WITH SIGNIFICANT COMMUNITIES OF INDIANS OUTSIDE INDIA
1994 or latest available date

Bangladesh communities over 1 percent of population

Afghanistan communities under 1 percent of population

Highest: Mauritius, 70% of total population; Guyana, 50%; Fiji, 48%;
Trinidad and Tobago, 41%

Sources: Kurian; Minority Rights Group; Parekh; UK Census of Population;
regional and national handbooks and yearbooks.

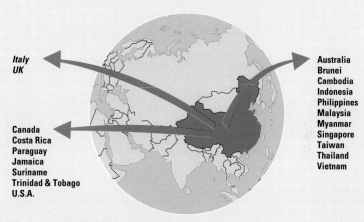

Italy
UK

Australia
Brunei
Cambodia
Indonesia
Philippines
Malaysia
Myanmar
Singapore
Taiwan
Thailand
Vietnam

Canada
Costa Rica
Paraguay
Jamaica
Suriname
Trinidad & Tobago
U.S.A.

THE CHINESE DIASPORA:
STATES WITH SIGNIFICANT COMMUNITIES OF OVERSEAS CHINESE
1994 or latest available date

Australia communities over 0.5 percent of population

Afghanistan communities under 0.5 percent of population

Highest: Singapore, 77% of total population; Malaysia, 30-35%; Brunei, 16%;
Thailand, 12%

Sources: *Independent*, London; Kurian; Minority Rights Group; UK Census of Population;
regional and national handbooks and yearbooks.

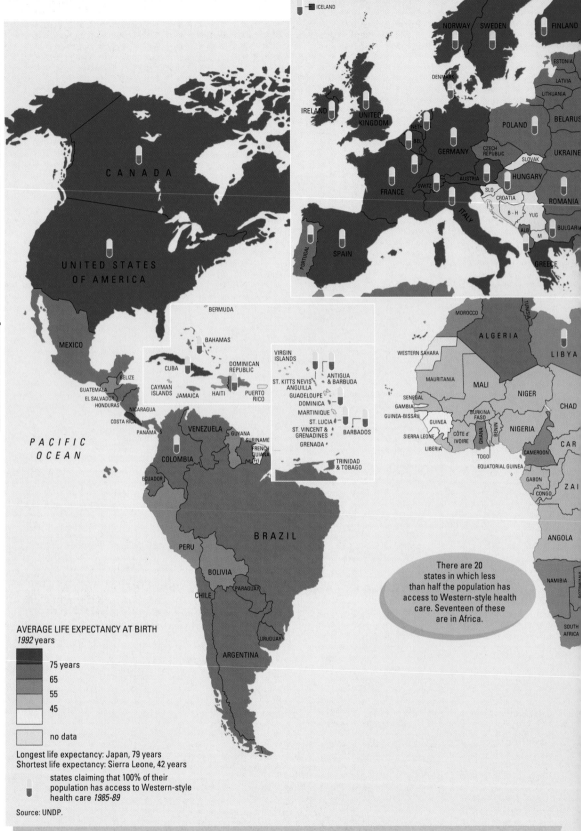

ICELAND

NORWAY SWEDEN FINLAND

ESTONIA
LATVIA
LITHUANIA

DENMARK

BELARUS

IRELAND UNITED KINGDOM POLAND

NETH
BEL GERMANY CZECH REPUBLIC UKRAINE
SLOVAK

FRANCE SWITZ AUSTRIA HUNGARY
SLO ROMANIA
CROATIA
B-H YUG
ITALY ALB M BULGARIA

PORTUGAL SPAIN
GREECE

C A N A D A

UNITED STATES
OF AMERICA

BERMUDA

BAHAMAS

MEXICO

CUBA
DOMINICAN REPUBLIC
CAYMAN ISLANDS JAMAICA HAITI PUERTO RICO

BELIZE
GUATEMALA
EL SALVADOR
HONDURAS
NICARAGUA

COSTA RICA
PANAMA

VIRGIN ISLANDS

ST. KITTS NEVIS ANGUILLA ANTIGUA & BARBUDA
GUADELOUPE
DOMINICA
MARTINIQUE
ST. LUCIA
ST. VINCENT & GRENADINES BARBADOS
GRENADA

TRINIDAD & TOBAGO

MOROCCO
TUNISIA

WESTERN SAHARA ALGERIA LIBYA

MAURITANIA MALI NIGER

SENEGAL
GAMBIA
GUINEA-BISSAU GUINEA BURKINA FASO CHAD

SIERRA LEONE CÔTE d'IVOIRE NIGERIA
LIBERIA GHANA BENIN
TOGO CAMEROON C A R

EQUATORIAL GUINEA

GABON Z A I
CONGO

PACIFIC OCEAN

VENEZUELA GUYANA
SURINAME
FRENCH GUIANA

COLOMBIA

ECUADOR

B R A Z I L

PERU

BOLIVIA

PARAGUAY
CHILE

ANGOLA

NAMIBIA

SOUTH AFRICA

There are 20 states in which less than half the population has access to Western-style health care. Seventeen of these are in Africa.

URUGUAY

ARGENTINA

AVERAGE LIFE EXPECTANCY AT BIRTH
1992 years

- 75 years
- 65
- 55
- 45
- no data

Longest life expectancy: Japan, 79 years
Shortest life expectancy: Sierra Leone, 42 years

states claiming that 100% of their population has access to Western-style health care *1985-89*

Source: UNDP.

People live longer when they have access to health care and clean water.

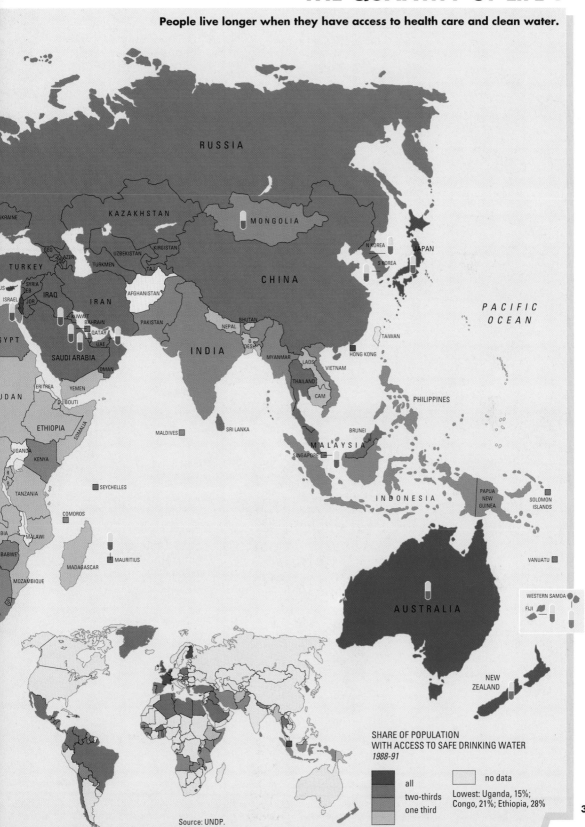

RUSSIA

KAZAKHSTAN

MONGOLIA

UKRAINE

GEO
AZER
TURKEY
UZBEKISTAN
KIRGISTAN
TURKMEN
TAJ

N KOREA

JAPAN

S KOREA

CHINA

SYRIA
DEB
IRAQ
JOR
ISRAEL

IRAN

AFGHANISTAN

PACIFIC
OCEAN

GYPT

KUWAIT
BAHRAIN
QATAR
UAE

PAKISTAN

BHUTAN
NEPAL
B
DESH

TAIWAN

SAUDI ARABIA

OMAN

INDIA

MYANMAR
LAOS

HONG KONG

SUDAN

ERITREA

YEMEN

DJIBOUTI

VIETNAM

THAILAND

CAM

PHILIPPINES

ETHIOPIA

SOMALIA

MALDIVES

SRI LANKA

BRUNEI

UGANDA

KENYA

R
B

SEYCHELLES

MALAYSIA

SINGAPORE

TANZANIA

COMOROS

INDONESIA

PAPUA
NEW
GUINEA

SOLOMON
ISLANDS

BIA

MALAWI

VANUATU

BABWE

MADAGASCAR

MAURITIUS

MOZAMBIQUE

S

AUSTRALIA

WESTERN SAMOA

FIJI

NEW
ZEALAND

SHARE OF POPULATION
WITH ACCESS TO SAFE DRINKING WATER
1988-91

all
two-thirds
one third

no data

Lowest: Uganda, 15%;
Congo, 21%; Ethiopia, 28%

Source: UNDP.

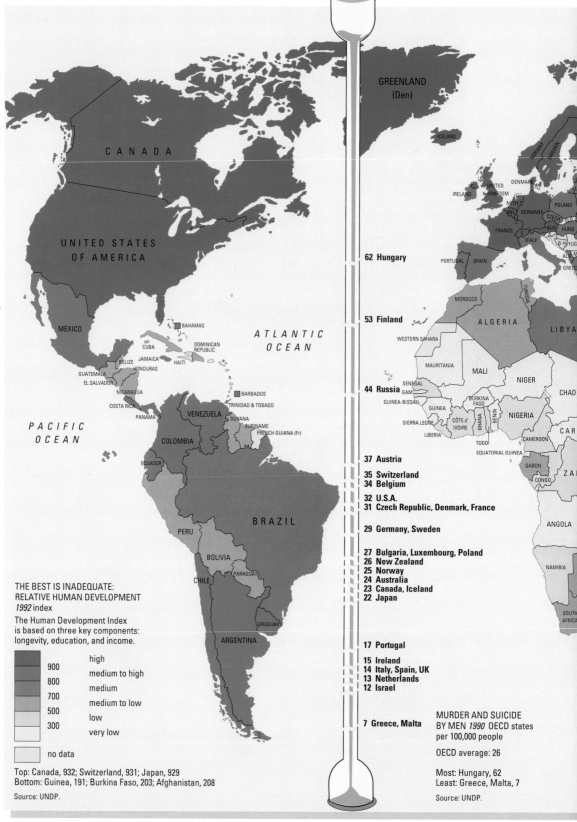

GREENLAND
(Den)

ICELAND

CANADA

UNITED STATES
OF AMERICA

62 Hungary

53 Finland

MEXICO

BAHAMAS

ATLANTIC
OCEAN

CUBA
DOMINICAN
REPUBLIC
BELIZE
JAMAICA
HAITI
GUATEMALA
HONDURAS
EL SALVADOR
NICARAGUA
BARBADOS
COSTA RICA
TRINIDAD & TOBAGO
PANAMA
VENEZUELA
GUYANA
SURINAME
FRENCH GUIANA (Fr)

PACIFIC
OCEAN

COLOMBIA

ECUADOR

BRAZIL

PERU

BOLIVIA

PARAGUAY

CHILE

URUGUAY

ARGENTINA

NORWAY SWEDEN

DENMARK
IRELAND
UNITED
KINGDOM
NETH
POLAND
BEL
GERMANY
CZECH
FRANCE
AUS HUNG
ITALY
B-H YUG
PORTUGAL SPAIN
ALB
GREECE

MOROCCO
TUNISIA

WESTERN SAHARA
ALGERIA
LIBYA

MAURITANIA
MALI
NIGER

SENEGAL
GAM
BURKINA
FASO
CHAD
GUINEA-BISSAU
GUINEA
BENIN
NIGERIA
SIERRA LEONE
CÔTE d'
IVOIRE
GHANA
CAR
LIBERIA
TOGO
CAMEROON
EQUATORIAL GUINEA
GABON
ZAI
CONGO

ANGOLA

NAMIBIA

SOUTH
AFRICA

44 Russia

37 Austria

35 Switzerland
34 Belgium

32 U.S.A.
31 Czech Republic, Denmark, France

29 Germany, Sweden

27 Bulgaria, Luxembourg, Poland
26 New Zealand
25 Norway
24 Australia
23 Canada, Iceland
22 Japan

17 Portugal

15 Ireland
14 Italy, Spain, UK
13 Netherlands
12 Israel

7 Greece, Malta

THE BEST IS INADEQUATE:
RELATIVE HUMAN DEVELOPMENT
1992 index

The Human Development Index
is based on three key components:
longevity, education, and income.

900	high
800	medium to high
700	medium
500	medium to low
300	low
	very low
	no data

Top: Canada, 932; Switzerland, 931; Japan, 929
Bottom: Guinea, 191; Burkina Faso, 203; Afghanistan, 208

Source: UNDP.

MURDER AND SUICIDE
BY MEN *1990* OECD states
per 100,000 people

OECD average: 26

Most: Hungary, 62
Least: Greece, Malta, 7

Source: UNDP.

In life, quality and quantity mostly march in step.

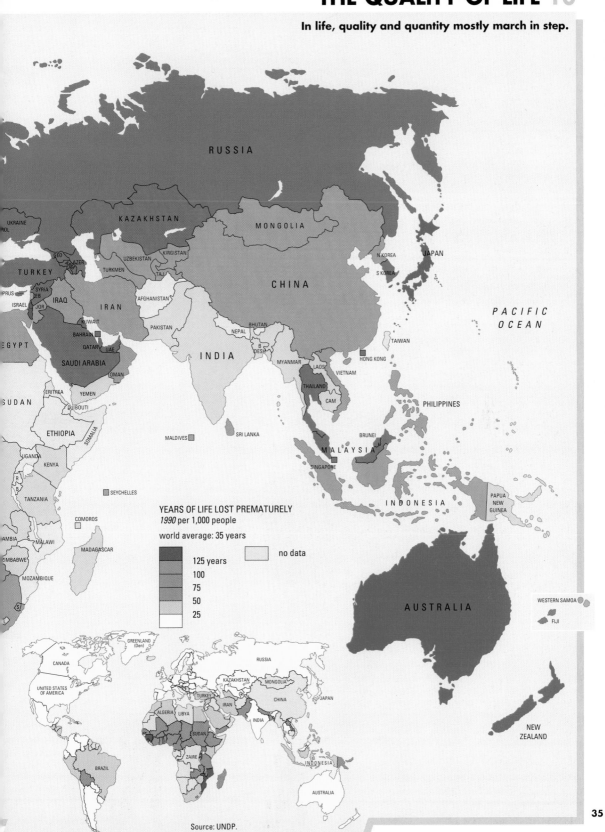

RUSSIA

KAZAKHSTAN

MONGOLIA

UKRAINE
MOL

GEO
AZER

UZBEKISTAN

KIRGISTAN

TURKMEN

TAJ

N.KOREA

JAPAN

S.KOREA

TURKEY

SYRIA
LEB

ISRAEL
JOR

IRAQ

IRAN

AFGHANISTAN

CYPRUS

CHINA

PACIFIC
OCEAN

KUWAIT

BAHRAIN

QATAR

UAE

EGYPT

SAUDI ARABIA

OMAN

PAKISTAN

NEPAL

BHUTAN

B
DESH

INDIA

MYANMAR

LAOS

VIETNAM

TAIWAN

HONG KONG

ERITREA

YEMEN

DJIBOUTI

SUDAN

ETHIOPIA

SOMALIA

THAILAND

CAM

PHILIPPINES

UGANDA

KENYA

MALDIVES

SRI LANKA

BRUNEI

R
B

TANZANIA

SEYCHELLES

MALAYSIA

SINGAPORE

COMOROS

INDONESIA

PAPUA
NEW
GUINEA

AMBIA

MALAWI

MADAGASCAR

ZIMBABWE

MOZAMBIQUE

S

YEARS OF LIFE LOST PREMATURELY
1990 per 1,000 people

world average: 35 years

125 years	no data
100	
75	
50	
25	

WESTERN SAMOA

FIJI

AUSTRALIA

GREENLAND
(Den)

CANADA

RUSSIA

UNITED STATES
OF AMERICA

KAZAKHSTAN

MONGOLIA

JAPAN

TURKEY

IRAN

CHINA

ALGERIA

LIBYA

INDIA

SUDAN

NEW
ZEALAND

BRAZIL

ZAIRE

INDONESIA

AUSTRALIA

Source: UNDP.

35

GREENLAND
(Den)

ICELAND

NORWAY
SWEDEN
FINL

DENMARK

IRELAND
UNITED
KINGDOM

NETH
BELG
GERMANY
FRANCE
ITALY

POLAND

CZECH
AUS HUNG
B-H SYUG
ALB
GREECE

PORTUGAL   SPAIN

C A N A D A

UNITED STATES
OF AMERICA

ATLANTIC
OCEAN

MEXICO

CUBA
JAMAICA
BELIZE   HAITI
HONDURAS
GUATEMALA
EL SALVADOR
NICARAGUA
COSTA RICA
PANAMA

DOMINICAN
REPUBLIC

TRINIDAD & TOBAGO

VENEZUELA
GUYANA
SURINAME
FRENCH GUIANA (Fr)

COLOMBIA

ECUADOR

B R A Z I L

PERU

BOLIVIA

PARAGUAY

CHILE

URUGUAY

ARGENTINA

PACIFIC
OCEAN

MOROCCO

WESTERN SAHARA

MAURITANIA

SENEGAL
GAMBIA
GUINEA-BISSAU
GUINEA
SIERRA LEONE
LIBERIA

CÔTE d'
IVOIRE

ALGERIA

MALI

BURKINA
FASO

GHANA
TOGO

TUNISIA

LIBYA

NIGER

BENIN

NIGERIA

EQUATORIAL GUINEA

CHAD

CAMEROON

CAR

GABON
CONGO

ZAIF

ANGOLA

NAMIBIA

BOTSWANA

SOUTH
AFRICA

INCOMES:
AVERAGE REAL
PURCHASING POWER
PER PERSON PER YEAR *1990*
international dollars

The international dollar is a measure
which attempts to gauge what states
could buy if they all shopped in a single
world supermarket.

	$16,000
	$8,000
	$4,000
	$2,000
	$1,000
	no data

Highest: U.S.A., $21,449
Lowest: Ethiopia, $369

Sources: UNDP; World Bank.

INEQUALITY:
DIFFERENCE IN INCOME BETWEEN THE RICHEST
10 PERCENT AND THE POOREST 20 PERCENT *1990*
selected states

Sources: UNDP; World Bank.

5 to 10 times

3 to 5 times

Canada, China,
Denmark, Ethiopia,
Finland, France,
Germany, Ghana, India,
Israel, Italy, Netherlands,
Norway, Pakistan,
Spain, Sweden,
Uganda, UK

Australia, Colombia,
Costa Rica, Hong Kong,
Lesotho, Malaysia,
Mexico, New Zealand,
Peru, Singapore,
Switzerland,
Thailand, U.S.A.,
Venezuela

The poor may inherit the earth some day. Meanwhile the rich are in possession.

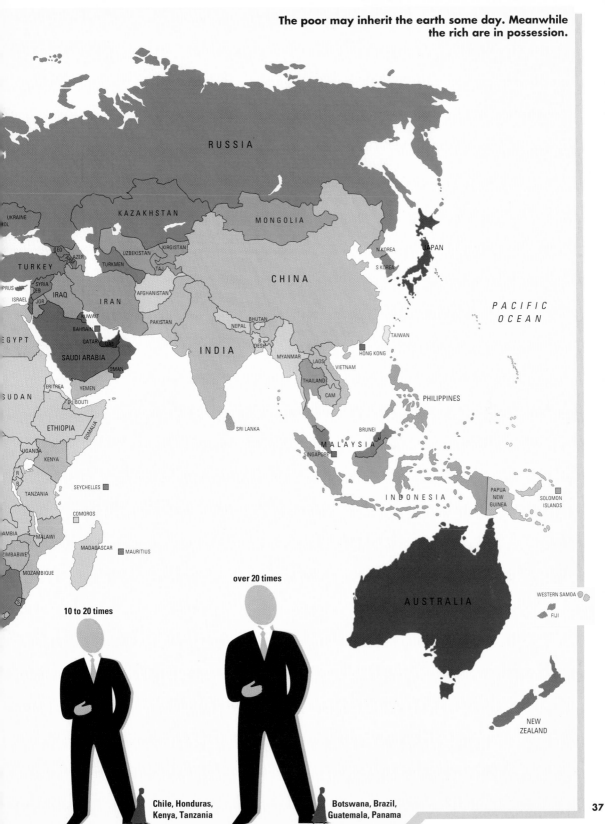

over 20 times

10 to 20 times

Chile, Honduras, Kenya, Tanzania

Botswana, Brazil, Guatemala, Panama

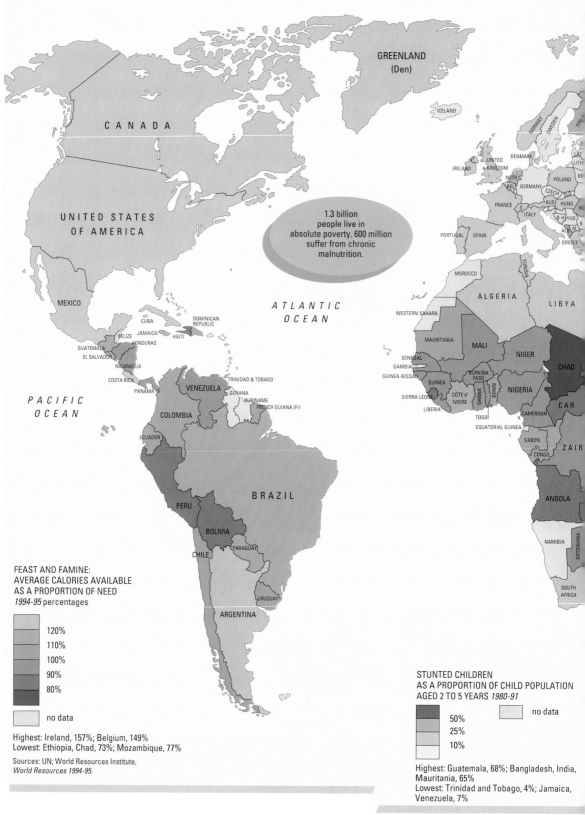

GREENLAND
(Den)

ICELAND

CANADA

UNITED STATES
OF AMERICA

NORWAY
SWEDEN
FINL
DENMARK
IRELAND
UNITED
KINGDOM
NETH
BEL
GERMANY
POLAND
CZECH
SLO
FRANCE
AUS
HUNG
ITALY
B-H/YUG
ALB
PORTUGAL   SPAIN
GREECE

1.3 billion
people live in
absolute poverty. 600 million
suffer from chronic
malnutrition.

ATLANTIC
OCEAN

MOROCCO
TUNISIA
WESTERN SAHARA
ALGERIA
LIBYA

MEXICO

CUBA
DOMINICAN
REPUBLIC
BELIZE   JAMAICA   HAITI
GUATEMALA   HONDURAS
EL SALVADOR
NICARAGUA
COSTA RICA
PANAMA

TRINIDAD & TOBAGO
VENEZUELA
GUYANA
SURINAME
FRENCH GUIANA (Fr)

COLOMBIA

ECUADOR

PACIFIC
OCEAN

MAURITANIA
MALI
NIGER
CHAD
SENEGAL
GAMBIA
GUINEA-BISSAU
BURKINA
FASO
GUINEA
SIERRA LEONE   CÔTE d'
IVOIRE
GHANA
BENIN
NIGERIA
LIBERIA
TOGO
CAMEROON
EQUATORIAL GUINEA
GABON
CONGO
ZAIR

CAR

BRAZIL

PERU

BOLIVIA

PARAGUAY

CHILE

ANGOLA

NAMIBIA
BOTSWANA

SOUTH
AFRICA

URUGUAY

ARGENTINA

FEAST AND FAMINE:
AVERAGE CALORIES AVAILABLE
AS A PROPORTION OF NEED
1994-95 percentages

- 120%
- 110%
- 100%
- 90%
- 80%
- no data

Highest: Ireland, 157%; Belgium, 149%
Lowest: Ethiopia, Chad, 73%; Mozambique, 77%

Sources: UN; World Resources Institute,
World Resources 1994-95.

STUNTED CHILDREN
AS A PROPORTION OF CHILD POPULATION
AGED 2 TO 5 YEARS *1980-91*

- 50%
- 25%
- 10%
- no data

Highest: Guatemala, 68%; Bangladesh, India,
Mauritania, 65%
Lowest: Trinidad and Tobago, 4%; Jamaica,
Venezuela, 7%

40 million people die each year from hunger-related diseases - the equivalent of over 300 jumbo-jet crashes a day with no survivors. Half the passengers are children.

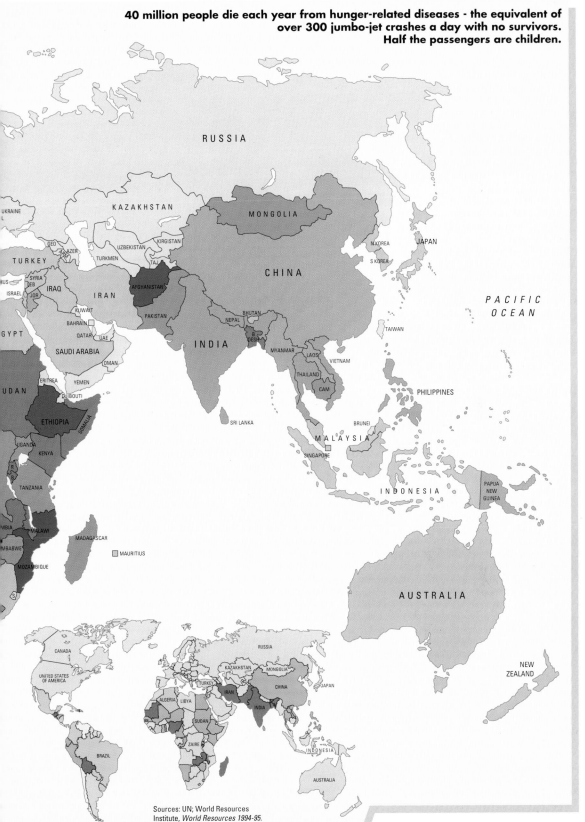

RUSSIA

KAZAKHSTAN

MONGOLIA

UKRAINE

GEO
AZER
TURKMEN
UZBEKISTAN
KIRGISTAN
TAJ.

N.KOREA

JAPAN

TURKEY

RUS
SYRIA
LEB
ISRAEL
JOR
IRAQ
IRAN
AFGHANISTAN

CHINA

S.KOREA

KUWAIT
BAHRAIN
QATAR
UAE

PAKISTAN

BHUTAN
NEPAL

TAIWAN

PACIFIC OCEAN

GYPT

SAUDI ARABIA

OMAN

INDIA

B. DESH

MYANMAR

LAOS

VIETNAM

ERITREA
YEMEN

THAILAND

UDAN

DJIBOUTI

CAM

PHILIPPINES

ETHIOPIA

SOMALIA

SRI LANKA

BRUNEI

UGANDA
KENYA

MALAYSIA

SINGAPORE

B

TANZANIA

INDONESIA

PAPUA NEW GUINEA

MBIA
MALAWI

MADAGASCAR

MBABWE

MAURITIUS

MOZAMBIQUE

AUSTRALIA

S

NEW ZEALAND

CANADA

RUSSIA

UNITED STATES OF AMERICA

KAZAKHSTAN

MONGOLIA

JAPAN

TURKEY

IRAN

CHINA

ALGERIA
LIBYA

INDIA

SUDAN

ZAIRE

BRAZIL

INDONESIA

AUSTRALIA

Sources: UN; World Resources
Institute, *World Resources 1994-95*.

INTERNATIONAL TELECOMMUNICATIONS
TRAFFIC OF INDIVIDUAL STATES COMPARED
WITH THAT OF THE U.S.A. *1992*
43 states (and U.S.A.) at least 44 millions MiTTs
(minutes of international telecoms traffic)

In 1992:
U.S.A.: 9,820 million MiTTs
Iceland: 44 million MiTTs

Source: Telegeography.

50% - 75% of U.S. traffic
France
UK

over 75% of U.S. traffic
Germany

U.S.A. 100%

Michael Kidron and Ronald Segal *The State of the World Atlas* 5th edition

GREENLAND
(Den)

U.S.A.

Sweden
Sweden

ICELAND

CANADA

UNITED
KINGDOM
IRELAND
UK
FRANCE

NORWAY
SWEDEN
FINLAND
E
DEN
Germany
N
BEL
GER POL
B CZ UKRAINE
A HUN ROM
B Y BULG
Germany

UNITED STATES
OF AMERICA

Germany
SPAIN
Germany
PORTUGAL
GREECE
Germany G AZER
TURKEY A TL
SYR
TUNISIA ISRAEL IRAQ IRAN

U.S.A.
MEXICO CUBA
DOMINICAN
REPUBLIC
BELIZE
HONDURAS HATI
GUATEMALA
EL SALVADOR NICARAGUA
COSTA RICA
PANAMA VENEZUELA GUYANA
SURINAME
U.S.A. FRENCH GUIANA (Fr)
COLOMBIA
ECUADOR
PERU
U.S.A.
U.S.A. BRAZIL
BOLIVIA
CHILE PAR

MOROCCO
WESTERN SAHARA ALGERIA LIBYA EGYPT
KUWAIT
QATAR
SAUDI
ARABIA
MAURITANIA MALI NIGER YEMEN
SENEGAL CHAD SUDAN
GAMBIA
GUINEA-BISSAU B F
GUINEA NIGERIA ETHIOPIA
CÔTE d' G B
SIERRA LEONE IVOIRE
LIBERIA CAR SOMALIA
TOGO CAM
EQUATORIAL GUINEA UG KENYA
GABON ZAIRE
CON TANZANIA
ANGOLA
ZAMBIA MALAWI
ZIM MADAGASCAR
NAMIBIA BOTS
MOZAMBIQUE
SOUTH
AFRICA

URU
ARGENTINA
Argentina

40

Some people see the world as their village. Most see their village as the world. They keep in contact with people they consider to be their neighbours.

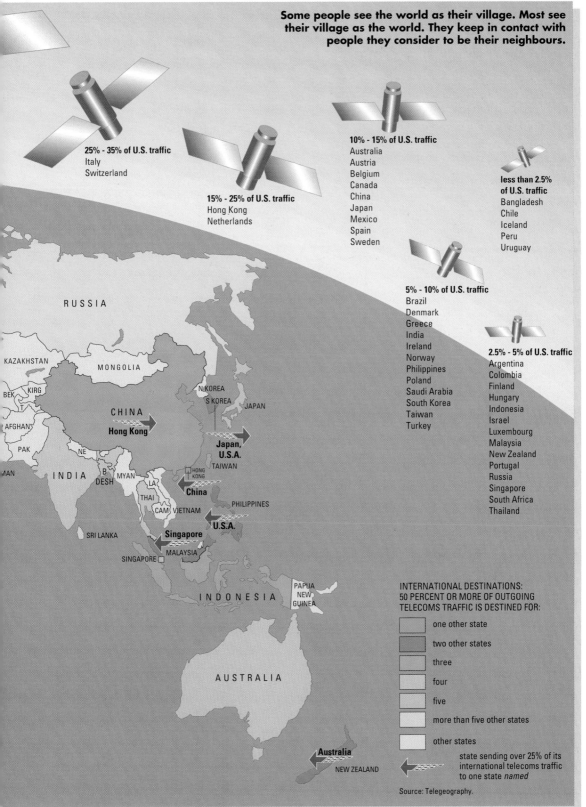

25% - 35% of U.S. traffic
Italy
Switzerland

15% - 25% of U.S. traffic
Hong Kong
Netherlands

10% - 15% of U.S. traffic
Australia
Austria
Belgium
Canada
China
Japan
Mexico
Spain
Sweden

less than 2.5% of U.S. traffic
Bangladesh
Chile
Iceland
Peru
Uruguay

5% - 10% of U.S. traffic
Brazil
Denmark
Greece
India
Ireland
Norway
Philippines
Poland
Saudi Arabia
South Korea
Taiwan
Turkey

2.5% - 5% of U.S. traffic
Argentina
Colombia
Finland
Hungary
Indonesia
Israel
Luxembourg
Malaysia
New Zealand
Portugal
Russia
Singapore
South Africa
Thailand

RUSSIA
KAZAKHSTAN
MONGOLIA
BEK
KIRG
AFGHAN
PAK
NE
INDIA
B
DESH
MYAN
MAN
SRI LANKA
LA
THAI
CAM VIETNAM
N.KOREA
S KOREA
JAPAN
TAIWAN
HONG KONG
PHILIPPINES
SINGAPORE
MALAYSIA
INDONESIA
PAPUA NEW GUINEA
AUSTRALIA
NEW ZEALAND

CHINA
Hong Kong
Japan, U.S.A.
China
U.S.A.
Singapore
Australia

INTERNATIONAL DESTINATIONS:
50 PERCENT OR MORE OF OUTGOING
TELECOMS TRAFFIC IS DESTINED FOR:

one other state
two other states
three
four
five
more than five other states
other states

state sending over 25% of its international telecoms traffic to one state *named*

Source: Telegeography.

41

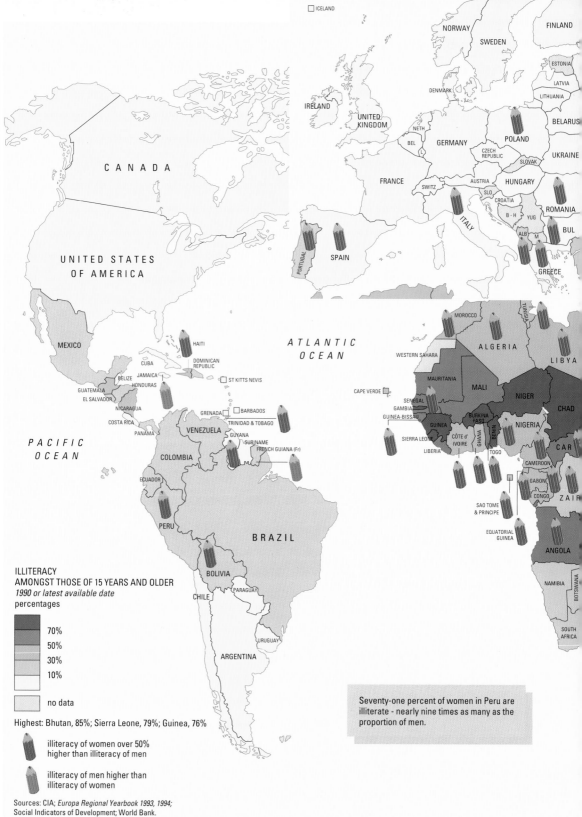

ICELAND

NORWAY
SWEDEN
FINLAND

DENMARK
ESTONIA
LATVIA
LITHUANIA

IRELAND

UNITED
KINGDOM
NETH
BEL
GERMANY
POLAND
BELARUS

CZECH
REPUBLIC
SLOVAK
UKRAINE

FRANCE
SWITZ
AUSTRIA
HUNGARY

SLO
CROATIA
ROMANIA

ITALY
B-H
YUG
BUL

ALB
M

GREECE

CANADA

UNITED STATES
OF AMERICA

PORTUGAL
SPAIN

ATLANTIC
OCEAN

MOROCCO
TUNISIA

WESTERN SAHARA
ALGERIA
LIBYA

MEXICO
HAITI

CUBA
DOMINICAN
REPUBLIC

BELIZE
JAMAICA
ST KITTS NEVIS

GUATEMALA
HONDURAS

EL SALVADOR

NICARAGUA
GRENADA
BARBADOS

COSTA RICA
TRINIDAD & TOBAGO

PANAMA

CAPE VERDE

MAURITANIA
MALI
NIGER
CHAD

SENEGAL
GAMBIA
BURKINA
FASO
NIGERIA

GUINEA-BISSAU
GUINEA
CÔTE d'
IVOIRE
BENIN
CAR

SIERRA LEONE
GHANA

LIBERIA
TOGO
CAMEROON

VENEZUELA
GUYANA
SURINAME
FRENCH GUIANA (Fr)

COLOMBIA

ECUADOR

SAO TOME
& PRINCIPE
GABON
CONGO
ZAIRE

PACIFIC
OCEAN

EQUATORIAL
GUINEA

PERU

BRAZIL
ANGOLA

BOLIVIA

NAMIBIA
BOTSWANA

PARAGUAY

CHILE

SOUTH
AFRICA

ILLITERACY
AMONGST THOSE OF 15 YEARS AND OLDER
1990 or latest available date
percentages

- 70%
- 50%
- 30%
- 10%

no data

Highest: Bhutan, 85%; Sierra Leone, 79%; Guinea, 76%

illiteracy of women over 50%
higher than illiteracy of men

illiteracy of men higher than
illiteracy of women

URUGUAY

ARGENTINA

Seventy-one percent of women in Peru are
illiterate - nearly nine times as many as the
proportion of men.

Sources: CIA; *Europa Regional Yearbook 1993, 1994;*
Social Indicators of Development; World Bank.

In most states, many people never learn to read or write. Educational neglect or other discrimination ensures that in all but a very few, more women than men are illiterate.

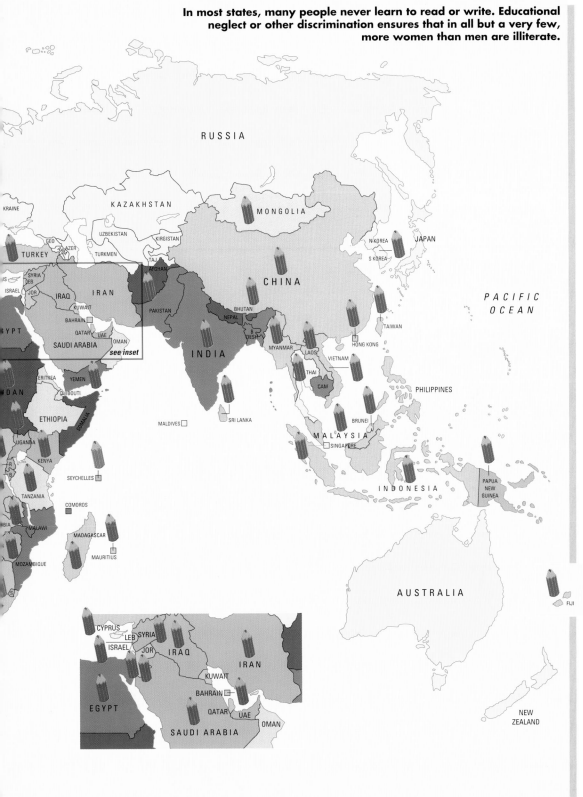

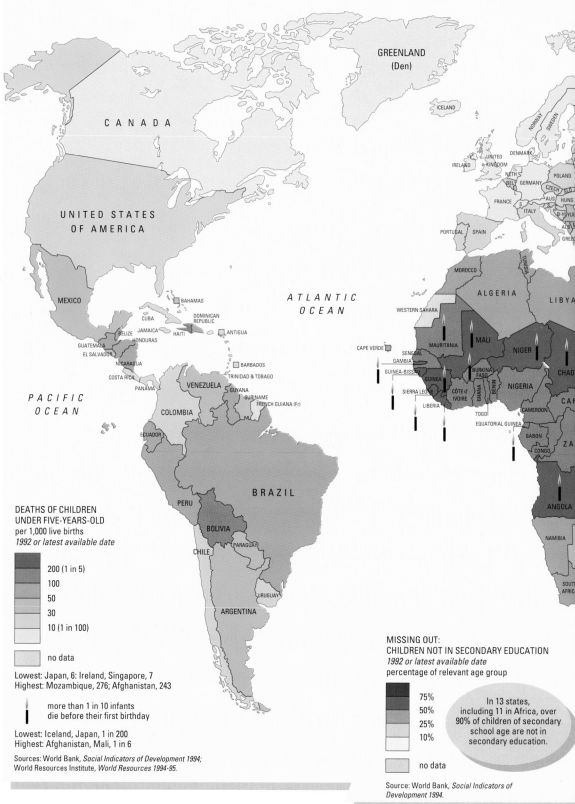

GREENLAND
(Den)

ICELAND

NORWAY
SWEDEN

DENMARK

IRELAND
UNITED
KINGDOM

NETH.
BELG. GERMANY

POLAND

CZECH
SLO

FRANCE

AUS HUNG

ITALY
B-HYUG

PORTUGAL SPAIN

ALB
GREEC

CANADA

UNITED STATES
OF AMERICA

MEXICO

BAHAMAS

CUBA
JAMAICA
BELIZE
HONDURAS
GUATEMALA
EL SALVADOR
NICARAGUA
COSTA RICA
PANAMA

DOMINICAN
REPUBLIC
HAITI

ANTIGUA

BARBADOS

TRINIDAD & TOBAGO

ATLANTIC
OCEAN

MOROCCO

WESTERN SAHARA

ALGERIA

LIBYA

TUNISIA

CAPE VERDE

MAURITANIA

MALI

NIGER

CHAD

SENEGAL
GAMBIA
GUINEA-BISSAU
SIERRA LEONE
LIBERIA

GUINEA

CÔTE d'
IVOIRE

BURKINA
FASO

GHANA
BENIN
TOGO

NIGERIA

CAMEROON

CAR

VENEZUELA

GUYANA
SURINAME
FRENCH GUIANA (Fr)

COLOMBIA

ECUADOR

PACIFIC
OCEAN

EQUATORIAL GUINEA

GABON
CONGO

ZA

BRAZIL

PERU

BOLIVIA

PARAGUAY

CHILE

URUGUAY

ARGENTINA

ANGOLA

NAMIBIA

SOUTH
AFRICA

DEATHS OF CHILDREN
UNDER FIVE-YEARS-OLD
per 1,000 live births
1992 or latest available date

- 200 (1 in 5)
- 100
- 50
- 30
- 10 (1 in 100)

- no data

Lowest: Japan, 6: Ireland, Singapore, 7
Highest: Mozambique, 276; Afghanistan, 243

more than 1 in 10 infants
die before their first birthday

Lowest: Iceland, Japan, 1 in 200
Highest: Afghanistan, Mali, 1 in 6

Sources: World Bank, *Social Indicators of Development 1994*;
World Resources Institute, *World Resources 1994-95*.

MISSING OUT:
CHILDREN NOT IN SECONDARY EDUCATION
1992 or latest available date
percentage of relevant age group

- 75%
- 50%
- 25%
- 10%

In 13 states,
including 11 in Africa, over
90% of children of secondary
school age are not in
secondary education.

- no data

Source: World Bank, *Social Indicators of
Development 1994.*

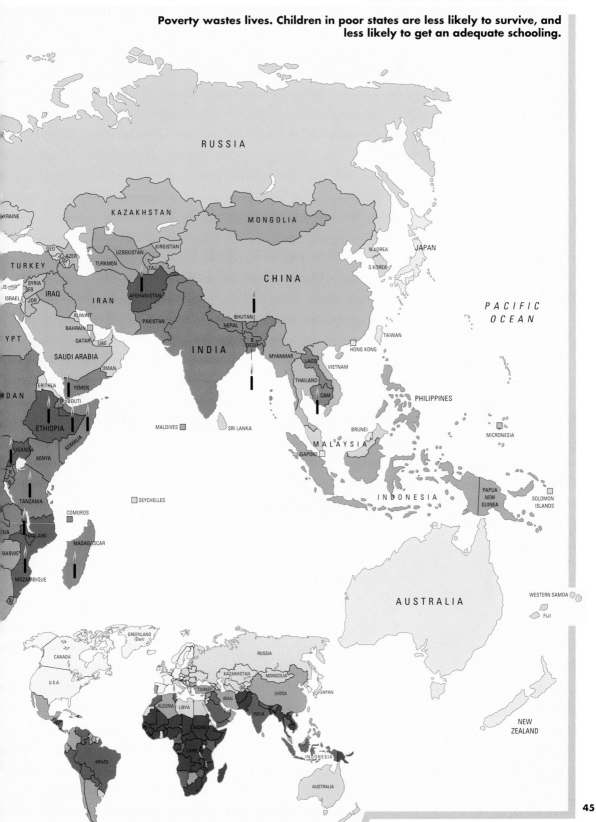

WASTING LIVES 15

Poverty wastes lives. Children in poor states are less likely to survive, and less likely to get an adequate schooling.

GREENLAND
(Den)

ICELAND

NORWAY
SWEDEN

CANADA

DENMARK
IRELAND   UNITED
KINGDOM
NETH
BELG GERMANY   POLAND
CZECH
FRANCE   AUS HUNG
ITALY   R-H YU
ALBA
PORTUGAL SPAIN   GREEC

UNITED STATES
OF AMERICA

MOROCCO   TUNISIA

ALGERIA
LIBYA
WESTERN SAHARA

MEXICO

ATLANTIC
OCEAN

MAURITANIA
MALI
NIGER
CHAD

CUBA
DOMINICAN
REPUBLIC

BELIZE JAMAICA   HAITI
HONDURAS
GUATEMALA
EL SALVADOR
NICARAGUA
COSTA RICA
PANAMA

SENEGAL
GAMBIA
GUINEA-BISSAU GUINEA
BURKINA
FASO
BENIN
NIGERIA

TRINIDAD & TOBAGO

SIERRA LEONE CÔTE d'
IVOIRE GHANA

CAR

VENEZUELA
GUYANA
SURINAME
FRENCH GUIANA (Fr)

LIBERIA
TOGO
EQUATORIAL GUINEA
CAMEROON

PACIFIC
OCEAN

COLOMBIA

GABON
CONGO
ZA

ECUADOR

ANGOLA

BRAZIL

PERU

NAMIBIA

BOLIVIA
PARAGUAY
CHILE

URUGUAY

SOUT
AFRIC

ARGENTINA

DEFICIT IN THE NUMBER OF WOMEN
AS A RESULT OF DISCRIMINATION
early 1990s percentages

Actual number of women compared with the
number that might have lived had they enjoyed
the same treatment they get in states of high
human development (for example, Canada,
Japan, Norway, Sweden, Switzerland).

- 10% deficit
- 5%
- 2.5%

surplus of women: in some cases
perhaps caused by war and/or homicide

least discriminating states:
number of women is 97% the number of men

no data

deficit of more than 500,000 women

Sources: UN; U.S. State Department.

If women worldwide enjoyed the same treatment as they do in Canada, Japan, Norway, Sweden and Switzerland, there would be 120 million more women alive. If they enjoyed fully-equal treatment with men, there would be many tens of millions more.

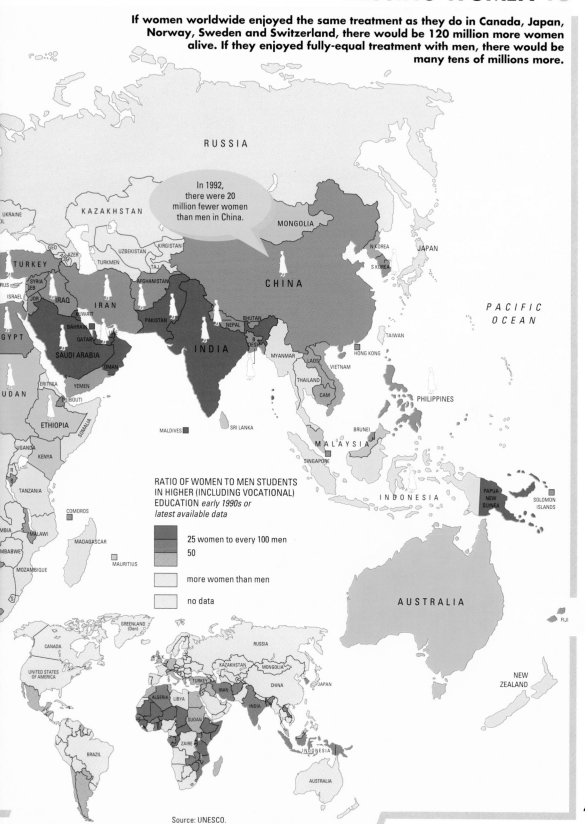

In 1992, there were 20 million fewer women than men in China.

RATIO OF WOMEN TO MEN STUDENTS IN HIGHER (INCLUDING VOCATIONAL) EDUCATION *early 1990s or latest available data*

- 25 women to every 100 men
- 50
- more women than men
- no data

Source: UNESCO.

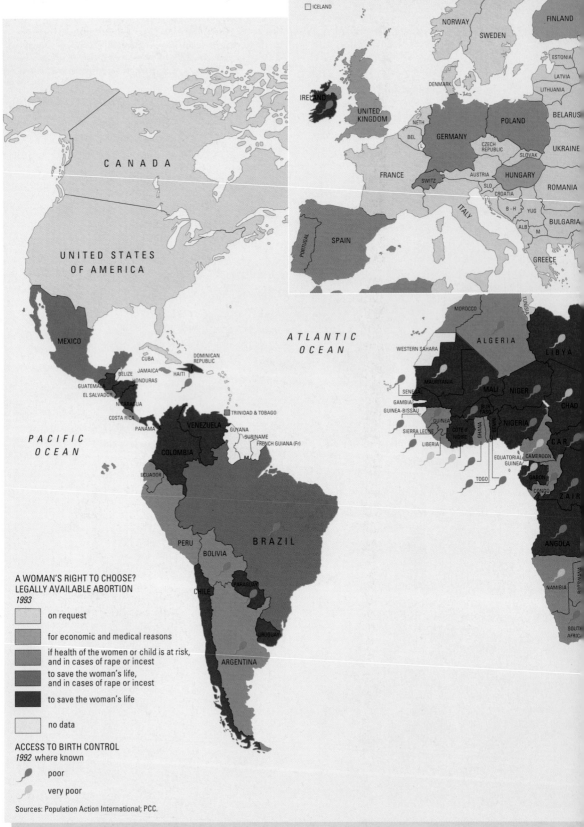

ICELAND

NORWAY

SWEDEN

FINLAND

DENMARK

ESTONIA

LATVIA

LITHUANIA

IRELAND

UNITED
KINGDOM

NETH

BEL

GERMANY

POLAND

BELARUS

UKRAINE

FRANCE

SWITZ

CZECH
REPUBLIC

SLOVAK

AUSTRIA

SLO

HUNGARY

ROMANIA

CROATIA

B - H

YUG

M

BULGARIA

PORTUGAL

SPAIN

ITALY

ALB

GREECE

CANADA

UNITED STATES
OF AMERICA

ATLANTIC
OCEAN

MEXICO

CUBA

JAMAICA

DOMINICAN
REPUBLIC

HAITI

BELIZE

HONDURAS

GUATEMALA

EL SALVADOR

NICARAGUA

COSTA RICA

PANAMA

TRINIDAD & TOBAGO

PACIFIC
OCEAN

VENEZUELA

GUYANA

SURINAME

FRENCH GUIANA (Fr)

COLOMBIA

ECUADOR

PERU

BRAZIL

BOLIVIA

MOROCCO

TUNISIA

WESTERN SAHARA

ALGERIA

LIBYA

MAURITANIA

MALI

NIGER

CHAD

SENEGAL

GAMBIA

GUINEA-BISSAU

BURK
FASO

GUINEA

NIGERIA

SIERRA LEONE

CÔTE d'
IVOIRE

GHANA

BENIN

LIBERIA

EQUATORIAL
GUINEA

CAMEROON

C A R

TOGO

GABON

CONGO

Z A I R

ANGOLA

NAMIBIA

BOTSWANA

SOUTH
AFRIC

PARAGUAY

CHILE

URUGUAY

ARGENTINA

A WOMAN'S RIGHT TO CHOOSE?
LEGALLY AVAILABLE ABORTION
1993

- on request
- for economic and medical reasons
- if health of the women or child is at risk, and in cases of rape or incest
- to save the woman's life, and in cases of rape or incest
- to save the woman's life
- no data

ACCESS TO BIRTH CONTROL
1992 where known

- poor
- very poor

Sources: Population Action International; PCC.

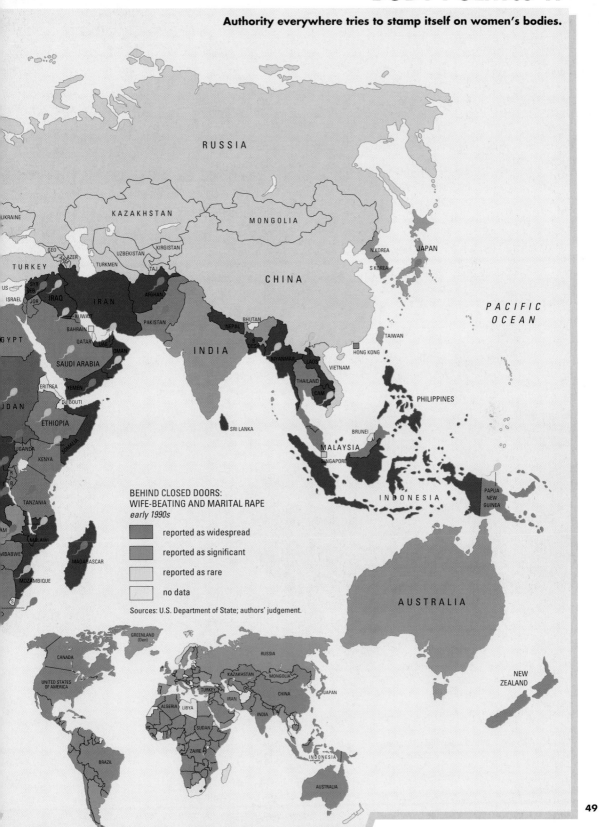

BEHIND CLOSED DOORS:
WIFE-BEATING AND MARITAL RAPE
early 1990s

- reported as widespread
- reported as significant
- reported as rare
- no data

Sources: U.S. Department of State; authors' judgement.

ICELAND

NORWAY

SWEDEN

FINLAND

DENMARK

ESTONIA

LATVIA

LITHUANIA

IRELAND

UNITED
KINGDOM

NETH

BELARUS

POLAND

BEL

GERMANY

UKRAIN

CZECH
REPUBLIC

SLOVAK

FRANCE

SWITZ

AUSTRIA

HUNGARY

SLO

CROATIA

ROMANIA

PORTUGAL

SPAIN

ITALY

B · H

YUG

ALB

M

BULGARIA

GREECE

CANADA

UNITED STATES
OF AMERICA

ATLANTIC
OCEAN

MOROCCO

TUNISIA

ALGERIA

LIBYA

WESTERN SAHARA

MEXICO

BAHAMAS

CUBA

HAITI

DOMINICAN
REPUBLIC

CAPE VERDE

MAURITANIA

MALI

NIGER

BELIZE

JAMAICA

GUATEMALA

HONDURAS

EL SALVADOR

NICARAGUA

GRENADA

TRINIDAD & TOBAGO

SENEGAL

GAMBIA

GUINEA-
BISSAU

GUINEA

BURKINA
FASO

NIGERIA

CHAD

COSTA RICA

PANAMA

VENEZUELA

GUYANA

SURINAME

FRENCH GUIANA (Fr)

SIERRA LEONE

LIBERIA

CÔTE d'
IVOIRE

GHANA

BENIN

TOGO

CAR

COLOMBIA

EQUATORIAL
GUINEA

CAMEROON

GABON

ECUADOR

SAO TOME
& PRINCIPE

CONGO

ZAI

9.8 U.S.A.

BRAZIL

PERU

BOLIVIA

ANGOLA

NAMIBIA

8.1 Denmark

7.4 Japan

6.9 New Zealand

6.2 Finland

CHILE

PARAGUAY

5.7 Hungary
5.5 Australia

SOUTH
AFRICA

4.9 Switzerland
4.7 Bulgaria

CHILD LABOUR:
MINIMUM AGE OF LEGAL EMPLOYMENT
1993

4.3 Austria, Belgium

URUGUAY

3.9 U.K.

ARGENTINA

12 years or no legal minimum

3.5 Germany

13 years

3.1 France

14 years

2.7 Canada

2.4 Poland

15 years

2.1 Netherlands

16 years or over

1.7 Norway

no data

INFANT DEATHS
FROM PRESUMED ABUSE
per 100,000 live births *1994*
selected countries

minimum age limit rarely enforced,
especially in agriculture and informal
economy

0.9 Sweden

0.4 Italy
0.2 Spain

Source: UNICEF.

Sources: International Confederation of Free Trade
Unions; U.S. State Department.

50

Children are widely exploited, not least for their labour.

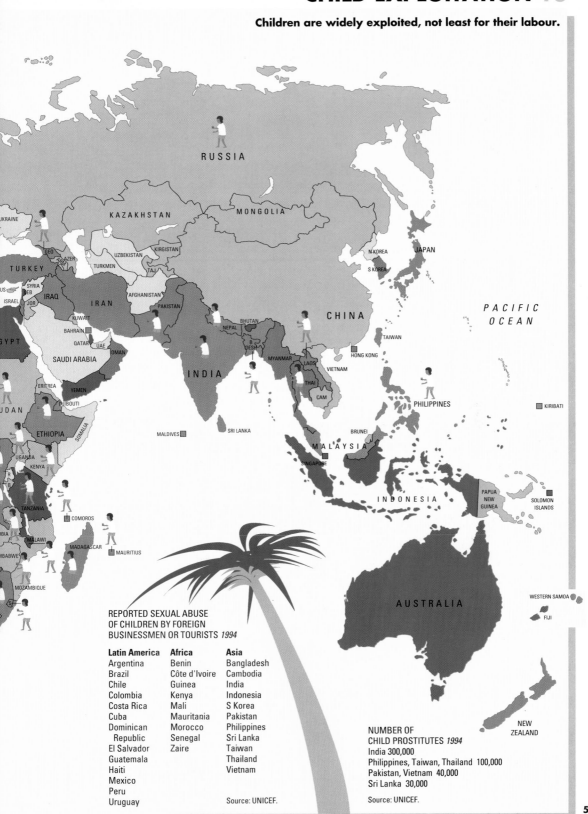

REPORTED SEXUAL ABUSE
OF CHILDREN BY FOREIGN
BUSINESSMEN OR TOURISTS *1994*

Latin America	Africa	Asia
Argentina	Benin	Bangladesh
Brazil	Côte d'Ivoire	Cambodia
Chile	Guinea	India
Colombia	Kenya	Indonesia
Costa Rica	Mali	S Korea
Cuba	Mauritania	Pakistan
Dominican	Morocco	Philippines
Republic	Senegal	Sri Lanka
El Salvador	Zaire	Taiwan
Guatemala		Thailand
Haiti		Vietnam
Mexico		
Peru		
Uruguay	Source: UNICEF.	

NUMBER OF
CHILD PROSTITUTES *1994*
India 300,000
Philippines, Taiwan, Thailand 100,000
Pakistan, Vietnam 40,000
Sri Lanka 30,000

Source: UNICEF.

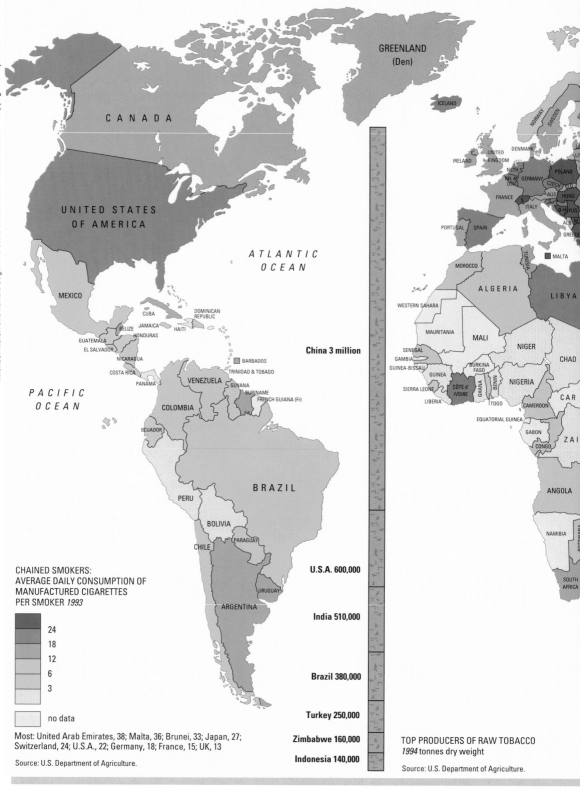

CANADA

GREENLAND
(Den)

ICELAND

UNITED STATES
OF AMERICA

ATLANTIC
OCEAN

MEXICO

CUBA
DOMINICAN
REPUBLIC
JAMAICA
BELIZE
HAITI
GUATEMALA
HONDURAS
EL SALVADOR

NICARAGUA

COSTA RICA
PANAMA

BARBADOS
TRINIDAD & TOBAGO

China 3 million

PACIFIC
OCEAN

VENEZUELA
GUYANA
SURINAME
FRENCH GUIANA (Fr)

COLOMBIA

ECUADOR

BRAZIL

PERU

BOLIVIA

PARAGUAY

CHILE

URUGUAY

ARGENTINA

NORWAY
SWEDEN
DENMARK

IRELAND

UNITED
KINGDOM

NETH.
BEL. &
LUX.
GERMANY
POLAND
CZECH
AUS.
HUNG.

FRANCE
ITALY

PORTUGAL
SPAIN

GREECE

MALTA

MOROCCO

TUNISIA

ALGERIA
LIBYA

WESTERN SAHARA

MAURITANIA
MALI
NIGER

CHAD

SENEGAL
GAMBIA
GUINEA-BISSAU
BURKINA
FASO
GUINEA
NIGERIA
SIERRA LEONE
CÔTE d'
IVOIRE
GHANA
BENIN
LIBERIA
TOGO
CAMEROON
CAR

EQUATORIAL GUINEA

GABON
ZAI

CONGO

ANGOLA

NAMIBIA
BOTSWANA

SOUTH
AFRICA

CHAINED SMOKERS:
AVERAGE DAILY CONSUMPTION OF
MANUFACTURED CIGARETTES
PER SMOKER *1993*

	24
	18
	12
	6
	3
	no data

Most: United Arab Emirates, 38; Malta, 36; Brunei, 33; Japan, 27;
Switzerland, 24; U.S.A., 22; Germany, 18; France, 15; UK, 13

Source: U.S. Department of Agriculture.

U.S.A. 600,000

India 510,000

Brazil 380,000

Turkey 250,000

Zimbabwe 160,000

Indonesia 140,000

TOP PRODUCERS OF RAW TOBACCO
1994 tonnes dry weight

Source: U.S. Department of Agriculture.

**Each day, between 1.3 and 1.4 billion people consume
more than 17,000 tonnes of tobacco.**

"One tonne of
tobacco = one million
cigarettes = one death"
Richard Peto, 1993

Exporters	Importers
Brazil 230,000	200,000 U.S.A.
U.S.A. 200,000	180,000 Germany
Zimbabwe 190,000	130,000 UK
Turkey 130,000	125,000 Russia
Greece 110,000	120,000 Japan
Italy 100,000	90,000 Netherlands
Malawi 100,000	50,000 Spain

TOP TRADERS IN RAW TOBACCO
1994 tonnes dry weight

Source: U.S. Department of Agriculture.

AIDS

REPORTED RATES OF AIDS
PER 100,000 POPULATION
1992 or latest available data

- 50 per 100,000
- 25
- 15
- 5
- 1

no cases reported

no data

Highest rate per 100,000: Bahamas, 104;
Namibia, 103; Congo, 85

more than 10,000 AIDS cases
reported to March 1994 *numbers given*

Source: WHO.

HIV-RELATED DEATHS OF ADULTS

- 2m by 1994
- 8m by 2000

HIV-INFECTED ADULTS

- 13-14m by 1994
- 20m by 2000

Source: WHO.

Map labels:

CANADA
UNITED STATES OF AMERICA 411,907
MEXICO 18,353
GUATEMALA
BELIZE
EL SALVADOR
HONDURAS
NICARAGUA
COSTA RICA
PANAMA
CUBA
JAMAICA
HAITI
DOMINICAN REPUBLIC
BAHAMAS
BERMUDA
ANGUILLA
ANTIGUA & BARBUDA
ST KITTS NEVIS
GUADELOUPE
DOMINICA
MARTINIQUE
ST. LUCIA
ST. VINCENT & GRENADINES
BARBADOS
GRENADA
TRINIDAD & TOBAGO
NETHERLANDS ANTILLES
COLOMBIA
VENEZUELA
GUYANA
SURINAME
FRENCH GUIANA (Fr)
ECUADOR
PERU
BRAZIL 49,312
BOLIVIA
PARAGUAY
CHILE
URUGUAY
ARGENTINA

PACIFIC OCEAN
ATLANTIC OCEAN

ICELAND
NORWAY
SWEDEN
FINLAND
IRELAND
UNITED KINGDOM
DENMARK
ESTONIA
LATVIA
LITHUANIA
NETH
BEL
GERMANY 11,179
POLAND
BELARUS
CZECH REPUBLIC
SLOVAK
UKRAINE
FRANCE 30,003
SWITZ
AUSTRIA
HUNGARY
SLO
CROATIA
B-H
YUG
ROMANIA
ITALY 21,770
ALB
M
BULGARIA
SPAIN 24,202
PORTUGAL
GREECE

MOROCCO
TUNISIA
ALGERIA
LIBYA
WESTERN SAHARA
MAURITANIA
MALI
NIGER
CHAD
CAPE VERDE
SENEGAL
GAMBIA
GUINEA-BISSAU
GUINEA
SIERRA LEONE
LIBERIA
BURKINA FASO
CÔTE d'IVOIRE
GHANA
BENIN
TOGO
NIGERIA
CAMEROON
CAR
EQUATORIAL GUINEA
SAO TOME & PRINCIPE
GABON
CONGO
ZAIRE 22,74
ANGOLA
NAMIBIA
BOTSWANA
SOUTH AFRICA
18,670
11,629

Michael Kidron and Ronald Segal *The State of the World Atlas* 5th edition   Copyright © Myriad Editions Limited

54

In the mid-1990s, AIDS and HIV infections are still increasing dramatically. Tuberculosis, for some years in retreat and supposedly curable, is increasing sharply.

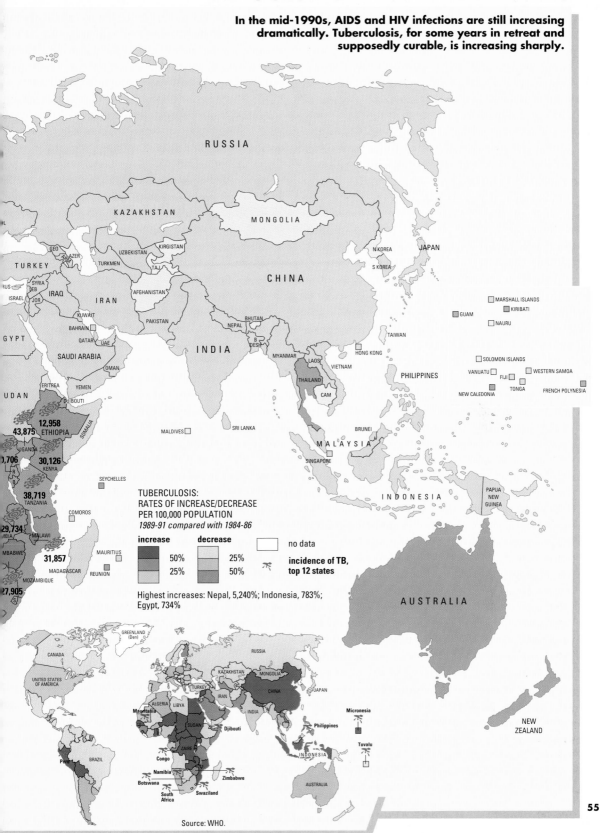

TUBERCULOSIS:
RATES OF INCREASE/DECREASE
PER 100,000 POPULATION
1989-91 compared with 1984-86

increase		decrease	
	50%		25%
	25%		50%

no data

incidence of TB, top 12 states

Highest increases: Nepal, 5,240%; Indonesia, 783%; Egypt, 734%

RUSSIA
KAZAKHSTAN
MONGOLIA
N.KOREA
JAPAN
S.KOREA
TURKEY
GEO
AZER
UZBEKISTAN
KIRGISTAN
TURKMEN
TAJ.
CHINA
SYRIA
DEB
IRAQ
IRAN
AFGHANISTAN
ISRAEL
JOR
TAIWAN
KUWAIT
PAKISTAN
BAHRAIN
NEPAL
BHUTAN
QATAR
UAE
SAUDI ARABIA
B DESH
INDIA
OMAN
MYANMAR
LAOS
HONG KONG
EGYPT
VIETNAM
ERITREA
YEMEN
PHILIPPINES
DJIBOUTI
THAILAND
UDAN
CAM
12,958
43,875 ETHIOPIA
SOMALIA
MALDIVES
SRI LANKA
BRUNEI
UGANDA
0,706
30,126 KENYA
MALAYSIA
SEYCHELLES
SINGAPORE
38,719 TANZANIA
COMOROS
INDONESIA
29,734
MBIA MALAWI
PAPUA NEW GUINEA
MBABWE
31,857
MAURITIUS
MADAGASCAR
REUNION
MOZAMBIQUE
27,905

MARSHALL ISLANDS
KIRIBATI
GUAM
NAURU
SOLOMON ISLANDS
VANUATU
WESTERN SAMOA
FIJI
NEW CALEDONIA
TONGA
FRENCH POLYNESIA

AUSTRALIA

GREENLAND (Den)
CANADA
RUSSIA
KAZAKHSTAN
MONGOLIA
UNITED STATES OF AMERICA
TURKEY
IRAN
JAPAN
ALGERIA LIBYA
Mauritania
INDIA
Micronesia
SUDAN
Djibouti
Philippines
ZAIRE
Tuvalu
Peru
BRAZIL
Congo
INDONESIA
Namibia
Zimbabwe
Botswana
AUSTRALIA
South Africa
Swaziland
NEW ZEALAND

Source: WHO.

The estimated value of the annual trade in futures' or derivatives' contracts ranges from U.S. $23 trillion (million million) to over six times that figure. It dwarfs the combined capital values of the world's export trade and its leading stock exchanges.

Reassurances that a market involving such enormous sums is fundamentally sound, since every contract to buy is balanced by one to sell, have proved increasingly hollow. Large industrial companies and banks have already been driven by their losses to the edge of extinction.

Source: press reports.

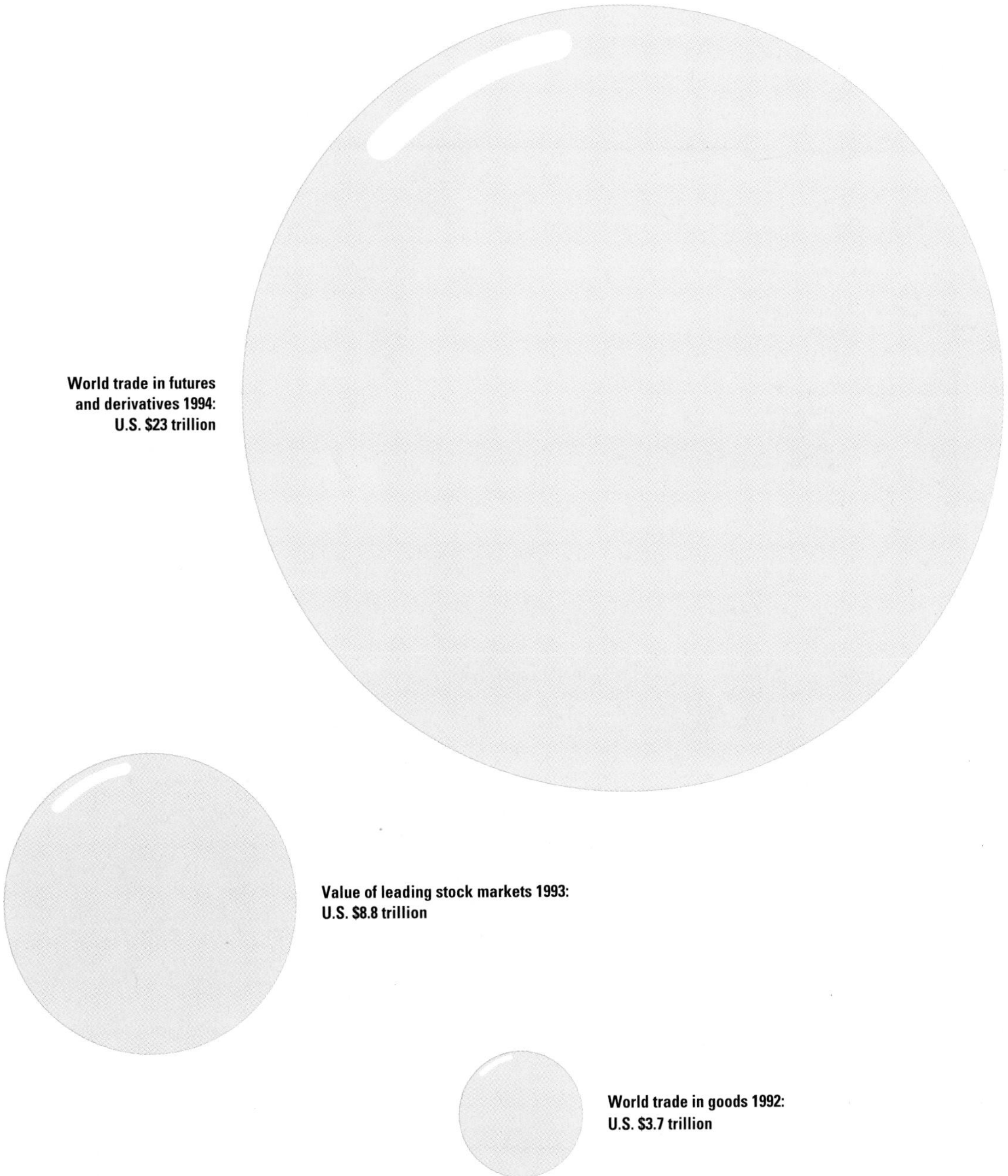

World trade in futures
and derivatives 1994:
U.S. $23 trillion

Value of leading stock markets 1993:
U.S. $8.8 trillion

World trade in goods 1992:
U.S. $3.7 trillion

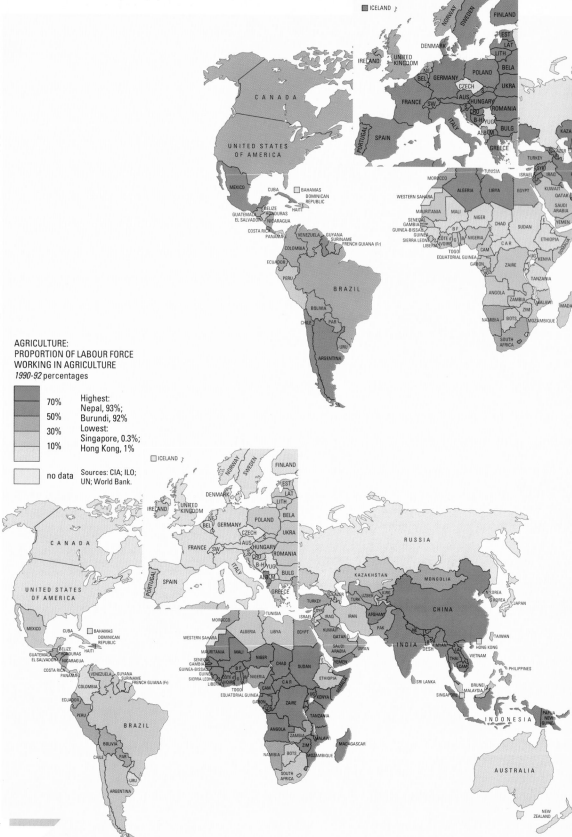

Michael Kidron and Ronald Segal *The State of the World Atlas* 5th edition   Copyright © Myriad Editions Limited

ICELAND

NORWAY SWEDEN FINLAND

DENMARK
IRELAND
UNITED KINGDOM
EST
LAT
LITH
NE BEL GERMANY POLAND BELA
FRANCE CZECH UKRA
AUS HUNGARY
SW CRO ROMANIA
B-H YUG
ITALY ALB BULG
PORTUGAL SPAIN GREECE

CANADA

UNITED STATES
OF AMERICA

MEXICO
CUBA
BAHAMAS
DOMINICAN REPUBLIC
BELIZE HAITI
GUATEMALA HONDURAS
EL SALVADOR NICARAGUA
COSTA RICA
PANAMA
VENEZUELA GUYANA
COLOMBIA SURINAME
FRENCH GUIANA (Fr)
ECUADOR
PERU

BRAZIL

BOLIVIA
PAR
CHILE
URU
ARGENTINA

TUNISIA
MOROCCO
WESTERN SAHARA ALGERIA LIBYA EGYPT
MAURITANIA MALI NIGER CHAD SUDAN
SENEGAL
GAMBIA
GUINEA-BISSAU B F NIGERIA
GUINEA CÔTE D'IV CAR
SIERRA LEONE
LIBERIA TOGO
EQUATORIAL GUINEA CAM
GABON ZAIRE
ANGOLA ZAMBIA
NAMIBIA BOTS
SOUTH AFRICA

TURKEY AZER
SYR IRAQ IRAN
ISRAEL KUWAIT QATAR
SAUDI ARABIA YEMEN
ETHIOPIA
UG KENYA
TANZANIA
MALAWI MADAGAS
ZIM
MOZAMBIQUE

AGRICULTURE:
PROPORTION OF LABOUR FORCE
WORKING IN AGRICULTURE
1990-92 percentages

70% Highest:
     Nepal, 93%;
50% Burundi, 92%
30% Lowest:
     Singapore, 0.3%;
10% Hong Kong, 1%

no data   Sources: CIA; ILO;
           UN; World Bank.

ICELAND

NORWAY SWEDEN FINLAND

DENMARK
IRELAND
UNITED KINGDOM
EST
LITH
NE POLAND BELA
BEL GERMANY
FRANCE CZECH UKRA
AUS HUNGARY
SW CRO ROMANIA
ITALY B-H YUG
PORTUGAL SPAIN ALB BULG
GREECE

RUSSIA

CANADA

UNITED STATES
OF AMERICA

KAZAKHSTAN
MONGOLIA
N KOREA
UZBEK KIRG S KOREA JAPAN
TURK
CHINA
TURKEY AZER AFGHAN
SYR IRAQ IRAN TAIWAN
ISRAEL PAK
KUWAIT NE
QATAR INDIA MYAN HONG KONG
SAUDI ARABIA BESH VIETNAM
OMAN THAI PHILIPPINES
YEMEN CAM
SRI LANKA
BRUNEI
SINGAPORE MALAYSIA

MEXICO
CUBA
BAHAMAS
DOMINICAN REPUBLIC
BELIZE HAITI
GUATEMALA HONDURAS
EL SALVADOR NICARAGUA
COSTA RICA
PANAMA VENEZUELA GUYANA
COLOMBIA SURINAME
FRENCH GUIANA (Fr)
ECUADOR
PERU

BRAZIL

BOLIVIA
PAR
CHILE
URU
ARGENTINA

MOROCCO TUNISIA
WESTERN SAHARA ALGERIA LIBYA EGYPT
MAURITANIA MALI NIGER CHAD SUDAN
SENEGAL
GAMBIA
GUINEA-BISSAU B F
GUINEA NIGERIA
SIERRA LEONE CÔTE D'IV CAR
LIBERIA TOGO
EQUATORIAL GUINEA CAM
GABON ZAIRE UG KENYA
TANZANIA
ANGOLA ZAMBIA
MALAWI
NAMIBIA ZIM MADAGASCAR
BOTS MOZAMBIQUE
SOUTH AFRICA

INDONESIA

PAPUA NEW GUINEA

AUSTRALIA

NEW ZEALAND

The richer the country, the more likely
that people in it will work
away from the land.

INDUSTRY:
PROPORTION OF LABOUR FORCE
WORKING IN INDUSTRY
1990-92 percentages

35%
25%
15%
5%

Highest: Bulgaria, 47%;
Romania, 46%
Lowest: Guinea, Nepal, 1%;
Bhutan, Burundi,
Equatorial Guinea, Mali,
Rwanda, 2%

no data

Sources: CIA; ILO; UN;
World Bank.

SERVICES:
PROPORTION OF LABOUR FORCE
WORKING IN SERVICES
1990-92 percentages

70%
50%
30%
10%

Highest: Bahrain, 83%;
Barbados, 82%
Lowest: Burundi,
Nepal, 6%;
Mozambique,
Rwanda, 8%

no data

Sources: CIA; ILO; UN;
World Bank.

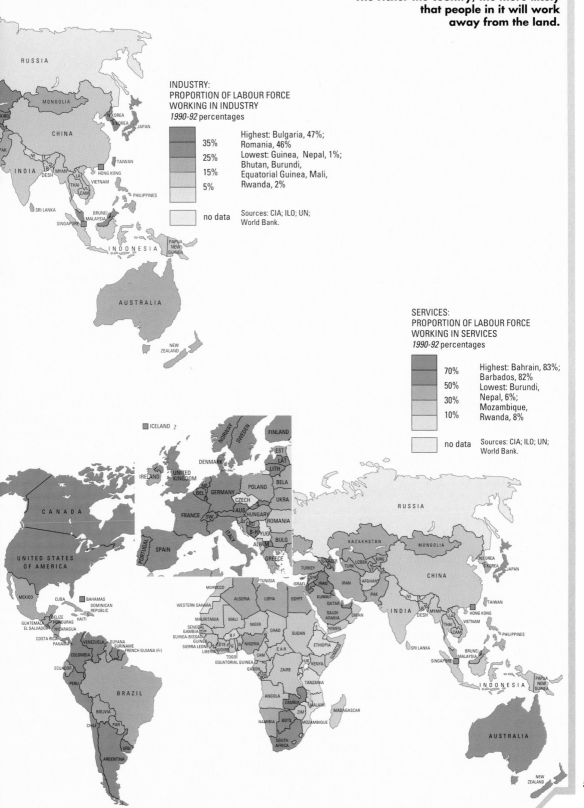

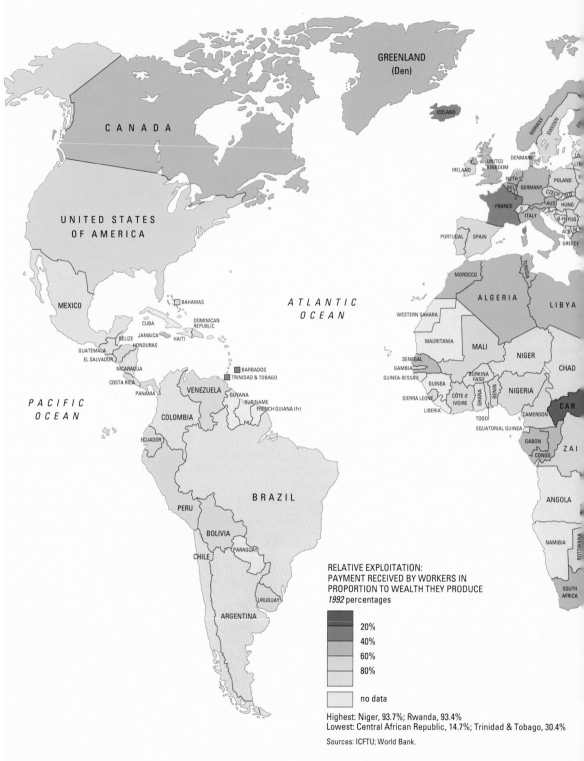

GREENLAND
(Den)

ICELAND

C A N A D A

NORWAY
SWEDEN
DENMARK
IRELAND
UNITED
KINGDOM
NETH.
BEL.
GERMANY
POLAND
CZECH
SLO
FRANCE
AUS.
HUNG
ITALY
B-H
SVUG
ALB.
PORTUGAL
SPAIN
GREECE

UNITED STATES
OF AMERICA

MEXICO

BAHAMAS

CUBA
BELIZE
JAMAICA
GUATEMALA
HONDURAS
HAITI
EL SALVADOR
NICARAGUA
COSTA RICA
PANAMA

DOMINICAN
REPUBLIC

ATLANTIC
OCEAN

BARBADOS
TRINIDAD & TOBAGO

MOROCCO
WESTERN SAHARA
ALGERIA
LIBYA
MAURITANIA
MALI
NIGER
CHAD
SENEGAL
GAMBIA
GUINEA-BISSAU
GUINEA
BURKINA
FASO
SIERRA LEONE
CÔTE d'
IVOIRE
GHANA
BENIN
NIGERIA
LIBERIA
TOGO
CAMEROON
CAR
EQUATORIAL GUINEA
GABON
ZAI
CONGO

VENEZUELA
GUYANA
SURINAME
FRENCH GUIANA (Fr)

COLOMBIA

ECUADOR

PACIFIC
OCEAN

BRAZIL

PERU

BOLIVIA

PARAGUAY

CHILE

URUGUAY

ARGENTINA

ANGOLA

NAMIBIA
BOTSWANA

SOUTH
AFRICA

RELATIVE EXPLOITATION:
PAYMENT RECEIVED BY WORKERS IN
PROPORTION TO WEALTH THEY PRODUCE
1992 percentages

20%
40%
60%
80%

no data

Highest: Niger, 93.7%; Rwanda, 93.4%
Lowest: Central African Republic, 14.7%; Trinidad & Tobago, 30.4%

Sources: ICFTU; World Bank.

Workers produce more wealth than they receive. The difference, after material costs have been met, is the value subtracted by capital - or the degree of exploitation.

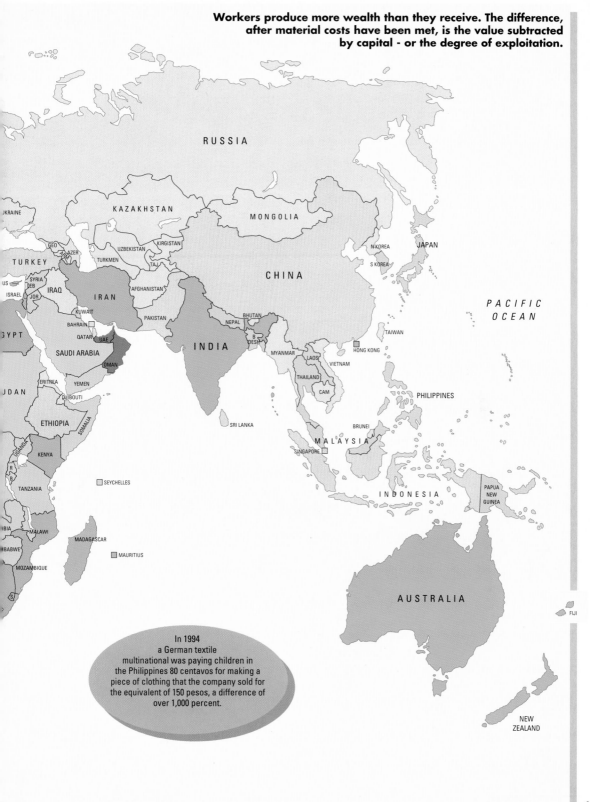

RUSSIA

UKRAINE

KAZAKHSTAN

MONGOLIA

GEO
AZER
UZBEKISTAN
KIRGISTAN

TURKEY
TURKMEN
TAJ.
N.KOREA
JAPAN

US
SYRIA
DEB
IRAQ
AFGHANISTAN
CHINA
S KOREA

ISRAEL
JOR
IRAN

KUWAIT
PAKISTAN

GYPT
BAHRAIN
QATAR
UAE
NEPAL
BHUTAN
TAIWAN

SAUDI ARABIA
OMAN
INDIA
B
DESH
HONG KONG

MYANMAR
LAOS

JDAN
ERITREA
YEMEN
THAILAND
VIETNAM

DJIBOUTI
CAM
PHILIPPINES

ETHIOPIA
SOMALIA
SRI LANKA
BRUNEI

UGANDA
KENYA
MALAYSIA

R
B
SINGAPORE

TANZANIA
SEYCHELLES
INDONESIA
PAPUA
NEW
GUINEA

IBIA
MALAWI
MADAGASCAR

BABWE
MAURITIUS

MOZAMBIQUE

PACIFIC
OCEAN

AUSTRALIA
FIJI

In 1994
a German textile
multinational was paying children in
the Philippines 80 centavos for making a
piece of clothing that the company sold for
the equivalent of 150 pesos, a difference of
over 1,000 percent.

NEW
ZEALAND

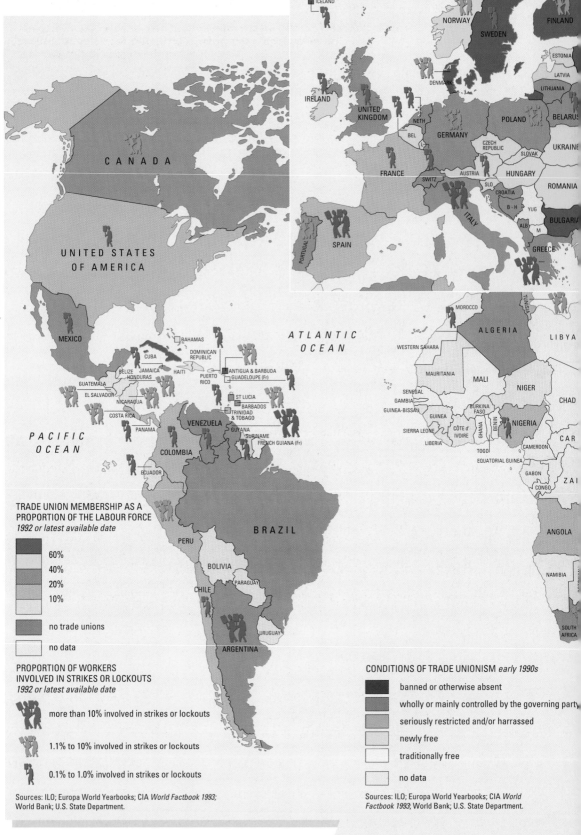

CANADA

UNITED STATES
OF AMERICA

ICELAND

NORWAY

SWEDEN

FINLAND

ESTONIA

LATVIA

LITHUANIA

DENMARK

IRELAND

UNITED
KINGDOM

NETH

BEL

GERMANY

POLAND

BELARUS

UKRAINE

CZECH
REPUBLIC

SLOVAK

FRANCE

SWITZ

AUSTRIA

HUNGARY

SLO

ROMANIA

CROATIA

B - H

YUG

BULGARIA

ALB

M

GREECE

PORTUGAL

SPAIN

ITALY

MEXICO

BAHAMAS

ATLANTIC
OCEAN

CUBA

DOMINICAN
REPUBLIC

BELIZE

JAMAICA

HAITI

GUATEMALA

HONDURAS

PUERTO
RICO

ANTIGUA & BARBUDA

GUADELOUPE (Fr)

EL SALVADOR

NICARAGUA

ST LUCIA

BARBADOS

COSTA RICA

TRINIDAD
& TOBAGO

PANAMA

VENEZUELA

GUYANA

PACIFIC
OCEAN

SURINAME

FRENCH GUIANA (Fr)

COLOMBIA

ECUADOR

MOROCCO

ALGERIA

LIBYA

WESTERN SAHARA

MAURITANIA

MALI

NIGER

CHAD

SENEGAL

GAMBIA

GUINEA-BISSAU

GUINEA

BURKINA
FASO

BENIN

NIGERIA

SIERRA LEONE

CÔTE d'
IVOIRE

GHANA

CAR

LIBERIA

TOGO

CAMEROON

EQUATORIAL GUINEA

GABON

ZAI

CONGO

TRADE UNION MEMBERSHIP AS A
PROPORTION OF THE LABOUR FORCE
1992 or latest available date

BRAZIL

PERU

60%

40%

20%

10%

no trade unions

no data

BOLIVIA

PARAGUAY

CHILE

ANGOLA

NAMIBIA

SOUTH
AFRICA

PROPORTION OF WORKERS
INVOLVED IN STRIKES OR LOCKOUTS
1992 or latest available date

URUGUAY

ARGENTINA

CONDITIONS OF TRADE UNIONISM *early 1990s*

more than 10% involved in strikes or lockouts

banned or otherwise absent

wholly or mainly controlled by the governing party

1.1% to 10% involved in strikes or lockouts

seriously restricted and/or harrassed

newly free

0.1% to 1.0% involved in strikes or lockouts

traditionally free

no data

Sources: ILO; Europa World Yearbooks; CIA *World Factbook 1993;*
World Bank; U.S. State Department.

Sources: ILO; Europa World Yearbooks; CIA *World
Factbook 1993;* World Bank; U.S. State Department.

TUNISIA

Free trade unions are an essential test of a democratic society. There are many societies, professing to be democratic, which do not pass it.

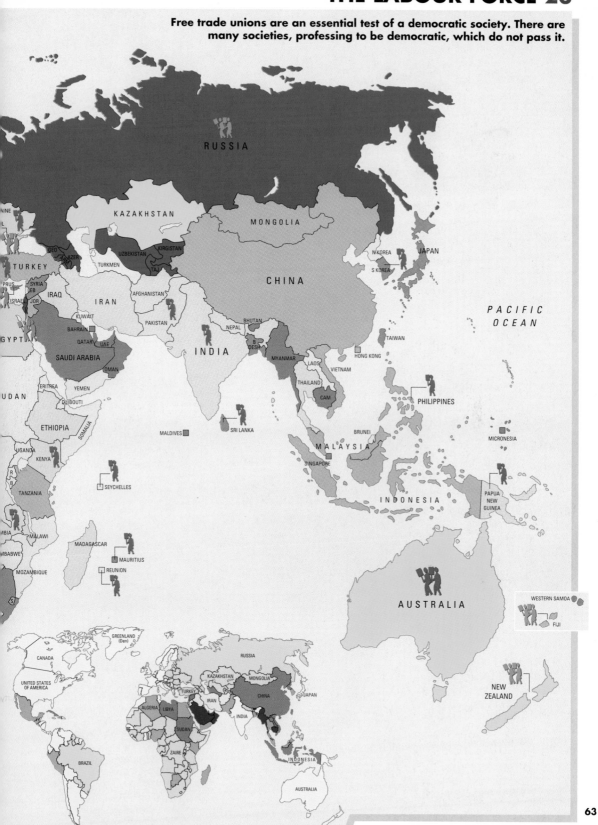

ICELAND

NORWAY SWEDEN

FINLAND

ESTONIA

LATVIA

LITHUANIA

IRELAND

DENMARK

BELARUS

UNITED
KINGDOM

NETH

POLAND

GERMANY

UKRAINE

BEL

CZECH
REPUBLIC

SLOVAK

FRANCE

AUSTRIA

HUNGARY

SWITZ

SLO

ROMANIA

CROATIA

B - H   YUG

ITALY

ALB   M

BULGARIA

PORTUGAL

SPAIN

GREECE

C A N A D A

UNITED STATES
OF AMERICA

There is no
paid maternity leave
in either the U.S.A. or
Australia.

*ATLANTIC
OCEAN*

MEXICO

CUBA

DOMINICAN
REPUBLIC

BELIZE   JAMAICA

HAITI

GUATEMALA   HONDURAS

EL SALVADOR

NICARAGUA

CAPE VERDE

COSTA RICA

PANAMA

VENEZUELA

TRINIDAD & TOBAGO

GUYANA

SURINAME

FRENCH GUIANA (Fr)

*PACIFIC
OCEAN*

COLOMBIA

ECUADOR

MOROCCO

ALGERIA

LIBYA

WESTERN SAHARA

MAURITANIA

MALI

NIGER

CHAD

SENEGAL

GAMBIA

GUINEA-BISSAU

BURKINA
FASO

NIGERIA

GUINEA

SIERRA LEONE

CÔTE d'
IVOIRE

GHANA

BENIN

CAR

LIBERIA

TOGO

CAMEROON

EQUATORIAL GUINEA

GABON

ZAI

CONGO

B R A Z I L

PERU

BOLIVIA

ANGOLA

PARAGUAY

CHILE

NAMIBIA

URUGUAY

SOUTH
AFRICA

ARGENTINA

WOMEN IN
THE LABOUR FORCE
1990-92 percentages

- 40% of labour force are women
- 30%
- 20%
- 10%

more women than men

no data

Highest: Cambodia, 56%; Rwanda, 54%
Lowest: Algeria, 4%; United Arab Emirates, 6%

PAID MATERNITY LEAVE
1991-92 rich states

over 20 weeks

up to 20 weeks

Between 1991
and 1994, women's share
of seats in the world's parliaments
fell from 13% to 10%. Less than 4% of
cabinet ministers and heads of
government were women.

A TOP JOB:
WOMEN'S SHARE
OF SEATS IN PARLIAMENT
1992 percentages

- 30%
- 20%
- 10%

no data

Highest: Finland, 38%;
Norway, 37%

Sources: *UN Human Development Report,* 1994; UNICEF,
The Progress of Nations, 1994; World Bank.

RISE OR FALL
1993 compared with 1987

● rise   ● fall

**In most states fewer women than men are in paid work.
Most of the top jobs still go to men.**

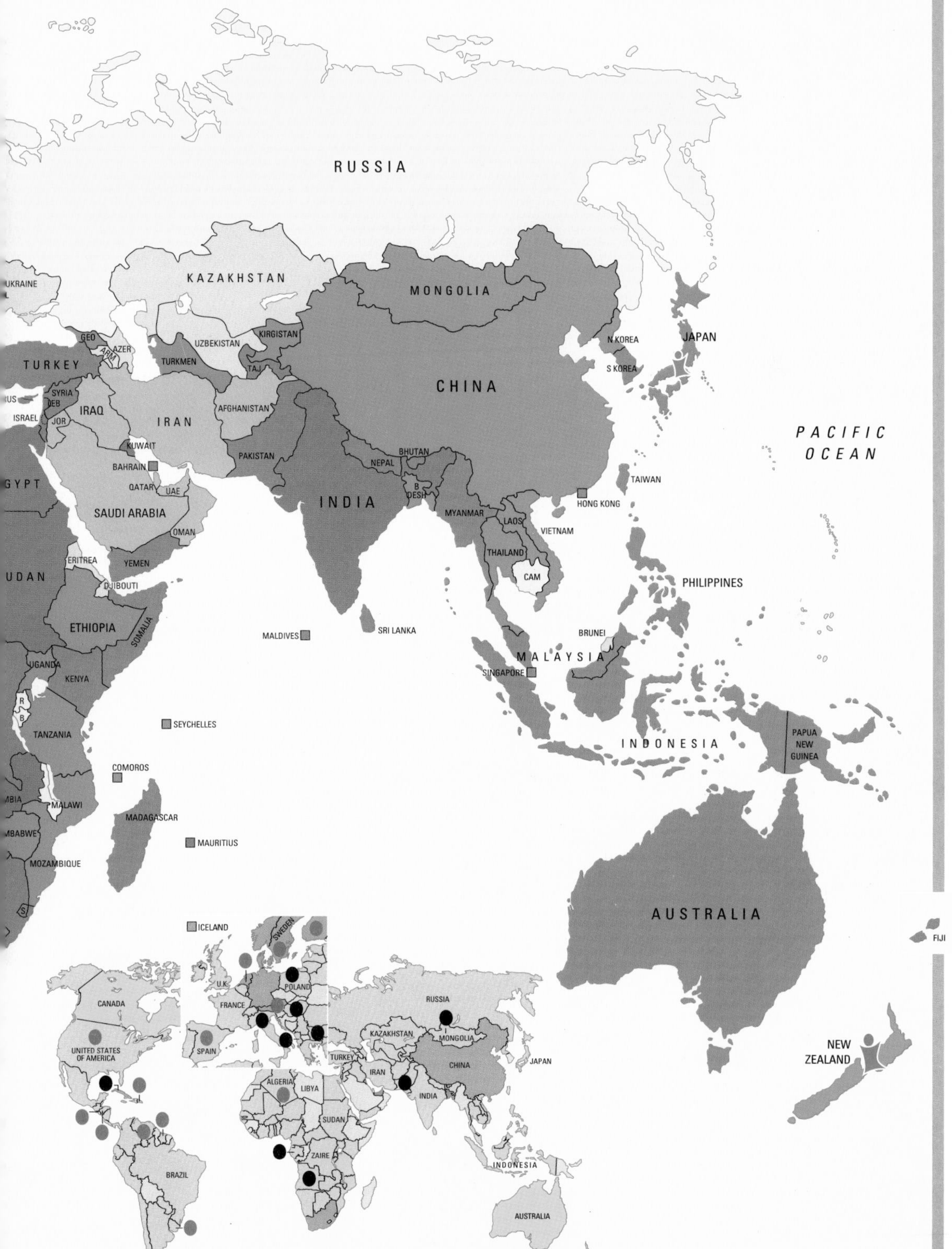

UKRAINE

RUSSIA

KAZAKHSTAN

MONGOLIA

GEO
ARM AZER
UZBEKISTAN KIRGISTAN
TURKEY TURKMEN TAJ.
RUS
SYRIA LEB
ISRAEL JOR IRAQ IRAN AFGHANISTAN
EGYPT KUWAIT PAKISTAN
BAHRAIN
QATAR UAE
SAUDI ARABIA OMAN
ERITREA YEMEN
SUDAN DJIBOUTI
ETHIOPIA

N KOREA
JAPAN
S KOREA

CHINA

PACIFIC
OCEAN

BHUTAN
NEPAL
B
DESH
INDIA MYANMAR LAOS
TAIWAN
HONG KONG
VIETNAM
THAILAND
CAM PHILIPPINES

SOMALIA
UGANDA
KENYA
R
B
TANZANIA

MALDIVES SRI LANKA

SEYCHELLES

COMOROS

BIA
MALAWI
MBABWE MADAGASCAR
MOZAMBIQUE MAURITIUS

BRUNEI
MALAYSIA
SINGAPORE

INDONESIA

PAPUA
NEW
GUINEA

AUSTRALIA

ICELAND SWEDEN

FIJI

U.K.
FRANCE POLAND
RUSSIA
CANADA KAZAKHSTAN MONGOLIA
SPAIN TURKEY CHINA JAPAN
UNITED STATES IRAN
OF AMERICA INDIA

NEW
ZEALAND

ALGERIA
LIBYA
SUDAN
ZAIRE
BRAZIL INDONESIA

AUSTRALIA

Sources: *UN Human Development Report*, 1994; UNICEF,
The Progress of Nations, 1994.

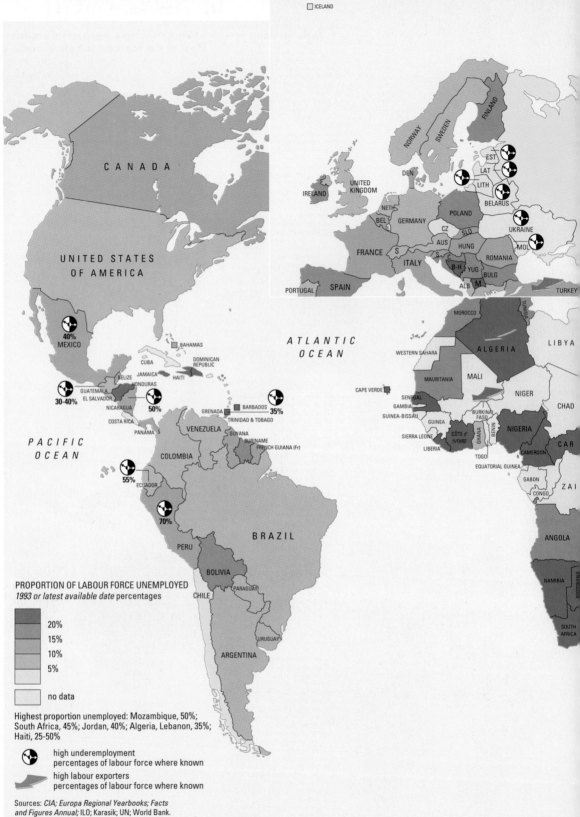

ICELAND

ICELAND

NORWAY SWEDEN FINLAND

IRELAND UNITED KINGDOM DEN

EST

LAT

LITH

BELARUS

NETH

BEL GERMANY POLAND

CZ SLO

AUS HUNG

UKRAINE

FRANCE S

ITALY

S
B-H YUG

C

ROMANIA

BULG

MOL

PORTUGAL SPAIN

ALB M

TURKEY

CANADA

UNITED STATES
OF AMERICA

40%
MEXICO

BAHAMAS

CUBA

DOMINICAN
REPUBLIC

ATLANTIC
OCEAN

BELIZE JAMAICA HAITI

GUATEMALA HONDURAS

30-40% EL SALVADOR

NICARAGUA

50%

COSTA RICA

PANAMA

GRENADA

TRINIDAD & TOBAGO

BARBADOS

35%

VENEZUELA GUYANA

PACIFIC
OCEAN

COLOMBIA

SURINAME

FRENCH GUIANA (Fr)

MOROCCO

TUNISIA

WESTERN SAHARA

ALGERIA

LIBYA

CAPE VERDE

MAURITANIA

MALI

NIGER

SENEGAL

CHAD

GAMBIA

GUINEA-BISSAU

BURKINA
FASO

GUINEA

SIERRA LEONE

CÔTE d'
IVOIRE

BENIN

NIGERIA

GHANA

55%
ECUADOR

70%

BRAZIL

PERU

BOLIVIA

LIBERIA

TOGO

CAMEROON

EQUATORIAL GUINEA

GABON

CAR

ZAI

CONGO

ANGOLA

NAMIBIA

PROPORTION OF LABOUR FORCE UNEMPLOYED
1993 or latest available date percentages

PARAGUAY

CHILE

SOUTH
AFRICA

20%

15%

10%

5%

no data

URUGUAY

ARGENTINA

Highest proportion unemployed: Mozambique, 50%;
South Africa, 45%; Jordan, 40%; Algeria, Lebanon, 35%;
Haiti, 25-50%

high underemployment
percentages of labour force where known

high labour exporters
percentages of labour force where known

Sources: *CIA; Europa Regional Yearbooks; Facts
and Figures Annual;* ILO; Karasik; UN; World Bank.

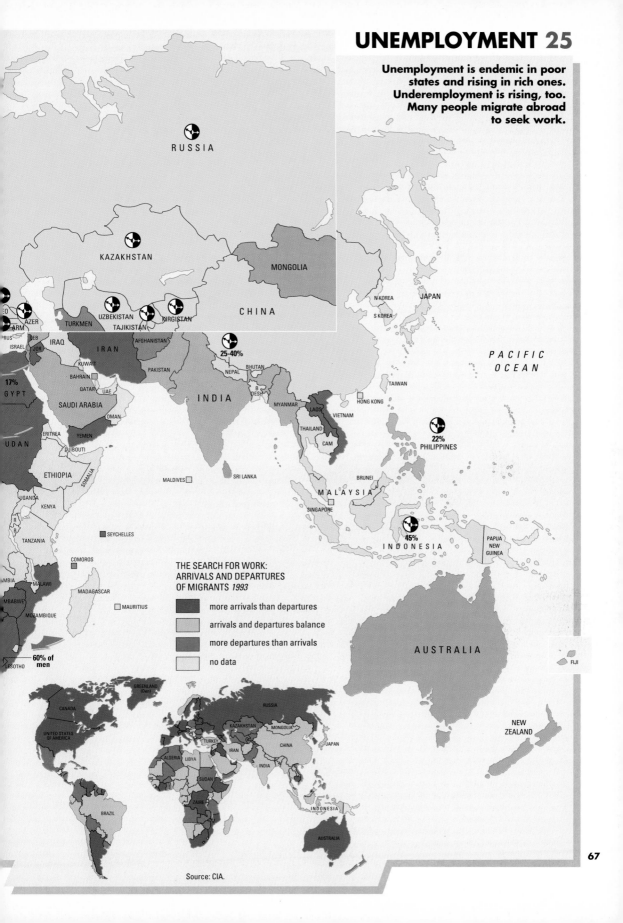

Unemployment is endemic in poor states and rising in rich ones. Underemployment is rising, too. Many people migrate abroad to seek work.

RUSSIA

KAZAKHSTAN

MONGOLIA

CHINA

ARM AZER

TURKMEN UZBEKISTAN KIRGISTAN

TAJIKISTAN

PRUS

ISRAEL JOR IRAQ IRAN AFGHANISTAN

17% EGYPT

KUWAIT

BAHRAIN
QATAR
UAE
SAUDI ARABIA
OMAN

ERITREA YEMEN
DJIBOUTI

UDAN

SOMALIA

ETHIOPIA

UGANDA
KENYA
R
B
TANZANIA

COMOROS

MBIA
MALAWI

MADAGASCAR

MBABWE

MOZAMBIQUE

60% of men
LESOTHO

PAKISTAN

NEPAL BHUTAN
B DESH

INDIA

MYANMAR

LAOS
THAILAND VIETNAM
CAM

MALDIVES

SRI LANKA

SEYCHELLES

MAURITIUS

25-40%

N KOREA JAPAN
S KOREA

TAIWAN

HONG KONG

22%
PHILIPPINES

BRUNEI

MALAYSIA

SINGAPORE

45%
INDONESIA

PACIFIC OCEAN

PAPUA NEW GUINEA

AUSTRALIA

FIJI

NEW ZEALAND

THE SEARCH FOR WORK: ARRIVALS AND DEPARTURES OF MIGRANTS *1993*

- more arrivals than departures
- arrivals and departures balance
- more departures than arrivals
- no data

CANADA
GREENLAND (Den)
RUSSIA
KAZAKHSTAN MONGOLIA
UNITED STATES OF AMERICA
TURKEY JAPAN
IRAN CHINA
ALGERIA LIBYA
INDIA
SUDAN
BRAZIL
ZAIRE
INDONESIA
AUSTRALIA

Source: CIA.

67

THE WORLD'S TOP 100 TRANSNATIONALS *1990*
Shares of total sales by headquarters state,
percentages

Total sales in 1990: $3,100 billion

Sources: UN, *World Investment Report:*
Transnational Corporations, 1993;
World Resources Institute, *World Resources 1994-95.*

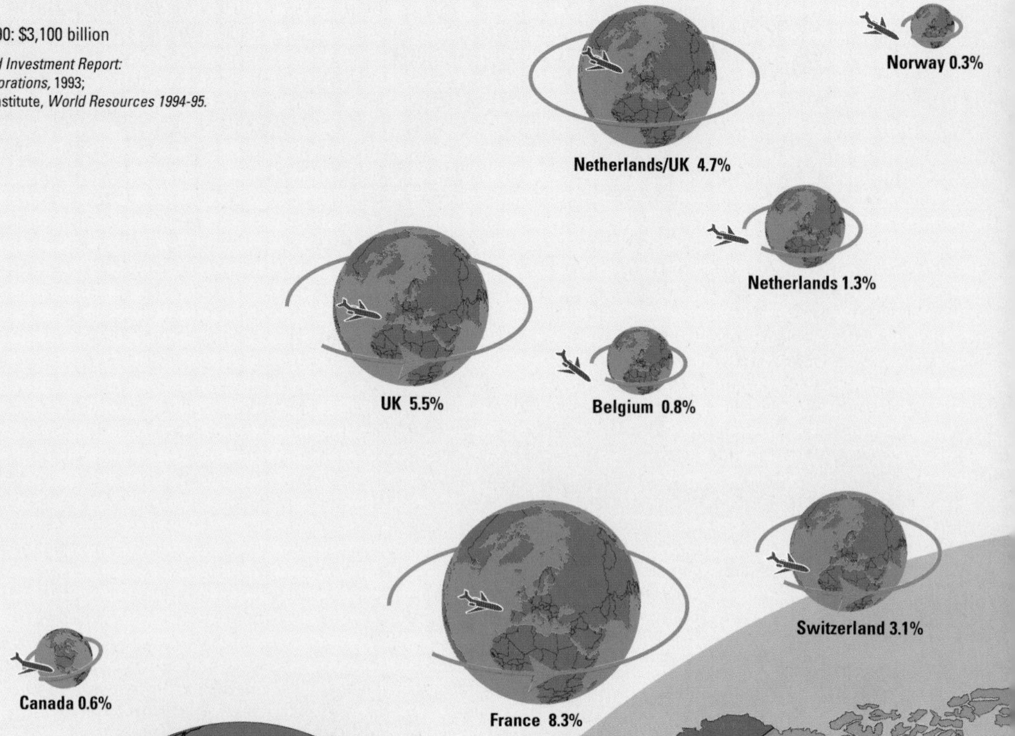

Norway 0.3%

Netherlands/UK 4.7%

Netherlands 1.3%

UK 5.5%

Belgium 0.8%

Switzerland 3.1%

Canada 0.6%

France 8.3%

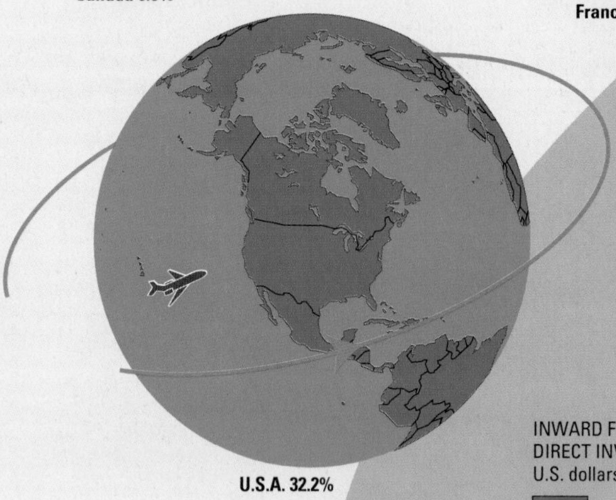

C A N A D A

UNITED STATES
OF AMERICA

MEXICO   CUBA
                              DOMINICAN
                              REPUBLIC
            BELIZE   HAITI
GUATEMALA HONDURAS
EL SALVADOR NICARAGUA
      COSTA RICA   VENEZUELA GUYANA
            PANAMA   SURINAME
                        COLOMBIA   FRENCH GUIANA (Fr)

                  ECUADOR

                  PERU   B R A Z I L

                          BOLIVIA

                    CHILE   PAR

                              URU

                        ARGENTINA

U.S.A. 32.2%

INWARD FLOWS OF FOREIGN
DIRECT INVESTMENT *1991*
U.S. dollars

$10 billion

$1 billion

$100 million

negative flow

other states

Source: UN, *World Investment Report:*
Transnational Corporations, 1993.

Transnationals invest in one another's backyards. Out of 37,000 transnationals worldwide, the top 100 have sales of $3,100 billion, almost half of which comes from abroad.

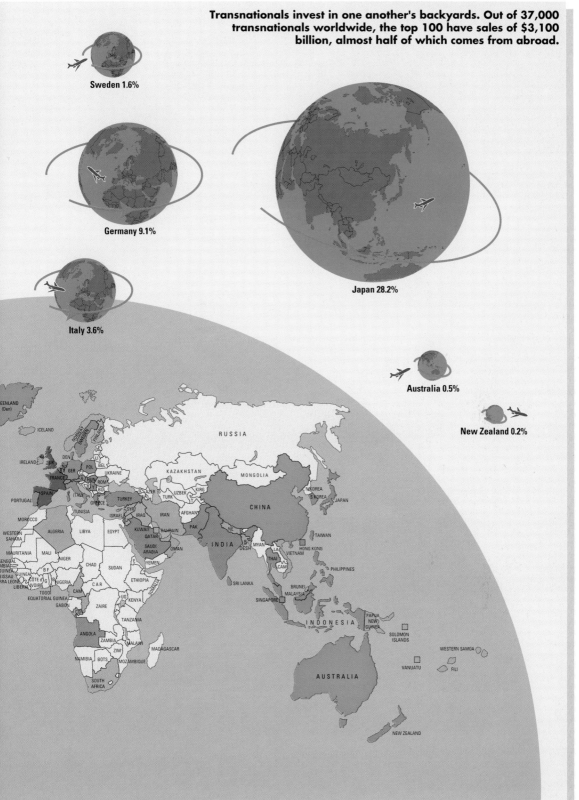

Sweden 1.6%

Germany 9.1%

Japan 28.2%

Italy 3.6%

Australia 0.5%

New Zealand 0.2%

GREENLAND (Den)

ICELAND

IRELAND

NORWAY SWEDEN FINLAND

DEN

GER POL BEL

FRANCE HUN ROM

SPAIN ITALY BUL

PORTUGAL GREECE TURKEY

TUNISIA

MOROCCO

WESTERN SAHARA

MAURITANIA MALI NIGER

SENEGAL

GAMBIA

GUINEA BISSAU GUINEA

SIERRA LEONE COTE D'IVOIRE

LIBERIA

ALGERIA LIBYA EGYPT

CHAD SUDAN

B F

NIGERIA

C A R ETHIOPIA

CAM

EQUATORIAL GUINEA

GABON CON ZAIRE

UG KENYA

TANZANIA

ANGOLA ZAMBIA MALAWI MADAGASCAR

NAMIBIA BOTS ZIM MOZAMBIQUE

SOUTH AFRICA

RUSSIA

KAZAKHSTAN

MONGOLIA

UKRAINE

AZER UZBEK KIRG

TURK

SYRI

ISRAEL IRAQ IRAN AFGHAN

KUWAIT

QATAR BAHRAIN PAK

SAUDI ARABIA OMAN

YEMEN

SRI LANKA

N.KOREA

S.KOREA JAPAN

CHINA

NE

INDIA DESH MYAN LA

BHU HONG KONG

THAI VIETNAM

CAM

TAIWAN

PHILIPPINES

BRUNEI MALAYSIA

SINGAPORE

INDONESIA

PAPUA NEW GUINEA

SOLOMON ISLANDS

WESTERN SAMOA

VANUATU FIJI

AUSTRALIA

NEW ZEALAND

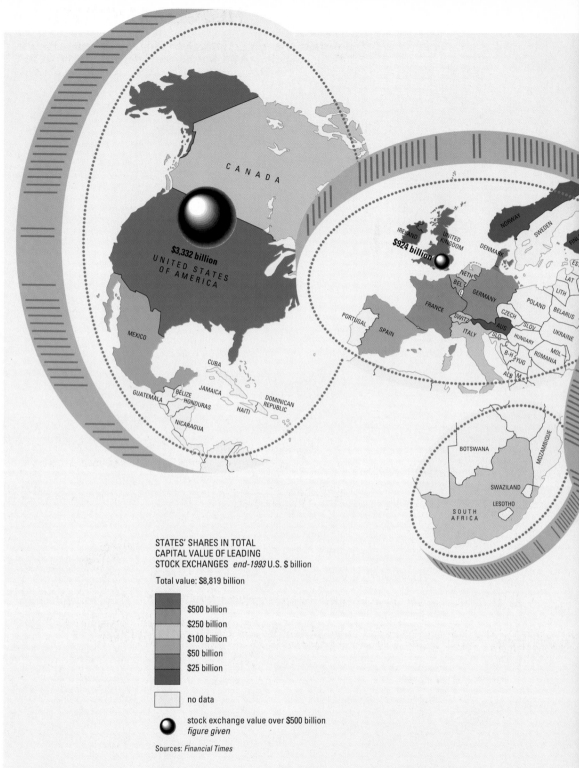

STATES' SHARES IN TOTAL
CAPITAL VALUE OF LEADING
STOCK EXCHANGES *end-1993* U.S. $ billion

Total value: $8,819 billion

- $500 billion
- $250 billion
- $100 billion
- $50 billion
- $25 billion

no data

stock exchange value over $500 billion
figure given

Sources: *Financial Times*

The line between investment and speculation has always been blurred. It is becoming more and more threateningly so.

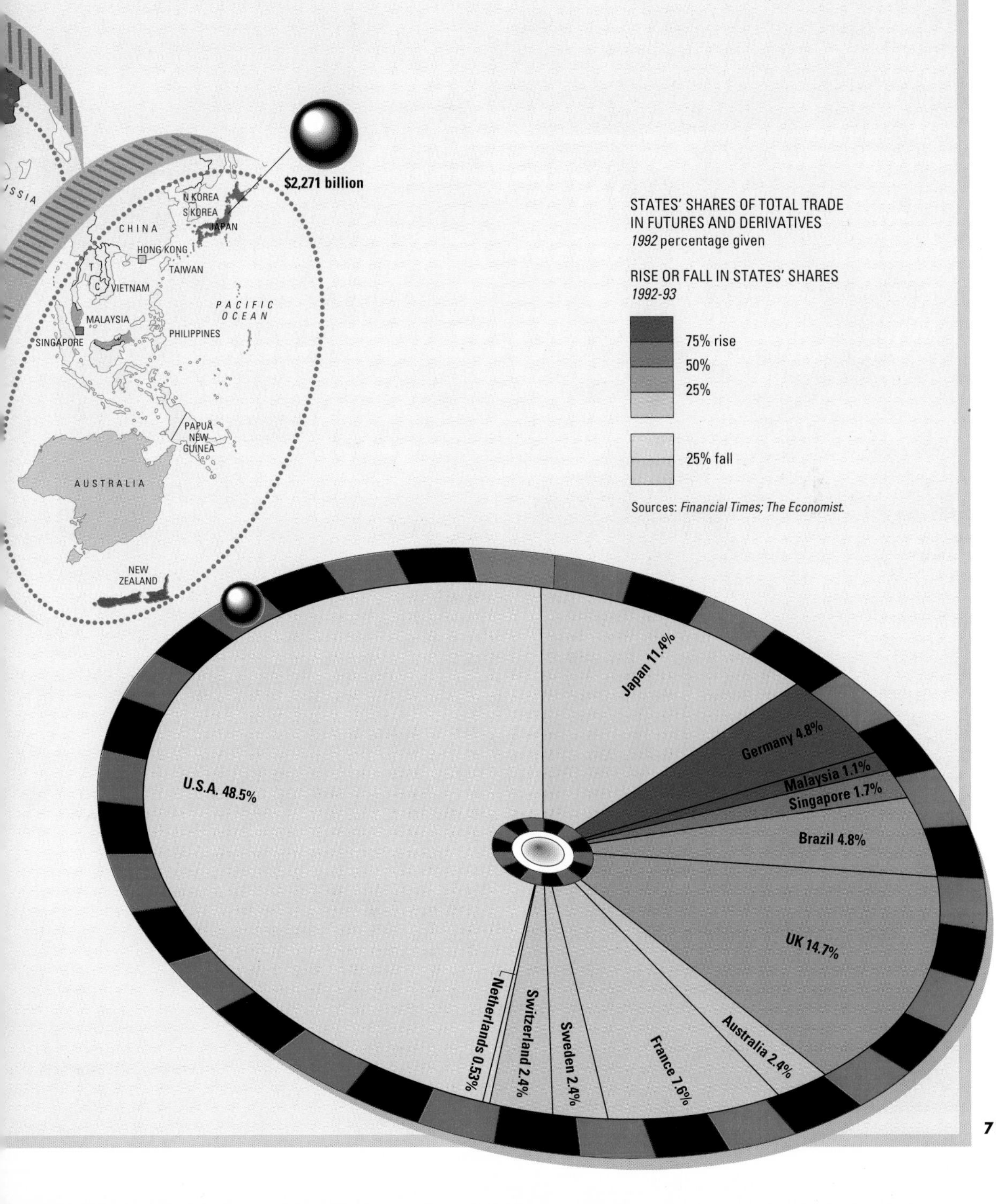

$2,271 billion

N KOREA
S KOREA
CHINA
JAPAN
HONG KONG
TAIWAN
T
C VIETNAM
PACIFIC
OCEAN
MALAYSIA
SINGAPORE
PHILIPPINES

PAPUA
NEW
GUINEA

AUSTRALIA

NEW
ZEALAND

RUSSIA

STATES' SHARES OF TOTAL TRADE
IN FUTURES AND DERIVATIVES
1992 percentage given

RISE OR FALL IN STATES' SHARES
1992-93

	75% rise
	50%
	25%
	25% fall

Sources: *Financial Times; The Economist.*

U.S.A. 48.5%

Japan 11.4%

Germany 4.8%

Malaysia 1.1%

Singapore 1.7%

Brazil 4.8%

UK 14.7%

Australia 2.4%

France 7.6%

Sweden 2.4%

Switzerland 2.4%

Netherlands 0.53%

ICELAND

NORWAY
SWEDEN
FINLAND

IRELAND
UNITED
KINGDOM
DENMARK
ESTONIA
LATVIA
LITHUANIA
BELARUS
POLAND
NETH
BEL
L
GERMANY
CZECH
REPUBLIC
SLOVAK
UKRAINE
FRANCE
SWITZ
AUSTRIA
HUNGARY
SLO
CROATIA
ROMANIA
B - H
YUG
BULGARIA
PORTUGAL
SPAIN
ITALY
ALB
M
GREECE

CANADA

UNITED STATES
OF AMERICA

HAITI

BERMUDA

MEXICO

BEL
GUATEMALA
EL SALVADOR
HONDURAS
NICARAGUA
COSTA RICA
PANAMA

BAHAMAS

CUBA

CAYMAN
ISLANDS
JAMAICA
HAITI

DOMINICAN
REPUBLIC

PUERTO
RICO

VIRGIN ISLANDS
ANGUILLA
ANTIGUA
& BARBUDA
ST. KITTS
NEVIS
MONTSERRAT
GUADELOUPE (Fr)
DOMINICA
MARTINIQUE (Fr)
ST. LUCIA
ST. VINCENT &
GRENADINES
BARBADOS
GRENADA

CAPE
VERDE

MOROCCO
TUNISIA

ALGERIA
LIBYA

WESTERN SAHARA

MAURITANIA
MALI
NIGER
CHAD

SENEGAL
GAMBIA
GUINEA-BISSAU
GUINEA
BURKINA
FASO
BENIN
NIGERIA
SIERRA LEONE
CÔTE d'
IVOIRE
GHANA
TOGO
CAMEROON
CAR
LIBERIA
EQUATORIAL GUINEA

PACIFIC
OCEAN

VENEZUELA
GUYANA
SURINAME
FRENCH
GUIANA
(Fr)

TRINIDAD
& TOBAGO

COLOMBIA

ECUADOR

PERU

BRAZIL

GABON
ZAIR
CONGO

ANGOLA

BOLIVIA

PARAGUAY

CHILE

NAMIBIA
BOTSWANA

SOUTH
AFRICA

ARGENTINA

URUGUAY

STATES' MAIN ROLE IN GLOBAL
PROHIBITED DRUG NETWORK
1993

drug producer

drug processor

conduit for drugs and/or processor
chemicals

drug consumer

no data or insignificant

business centres for top level negotiations

states whose role in the drug network
makes them essential to the industry

major drug money laundering sites

Source: U.S. State Department.

Most states play a part in the global market for prohibited drugs, a market almost as large as those for the major permitted drugs – nicotine and alcohol.

ANTI-DRUG ENFORCEMENT BODIES
(POLICE, CUSTOMS, JUDICIARY)
TAINTED BY DRUG MONEY *1993*

- corruption at senior level
- corruption at street level
- no data or insignificant

Sources: U.S. State Department; press reports.

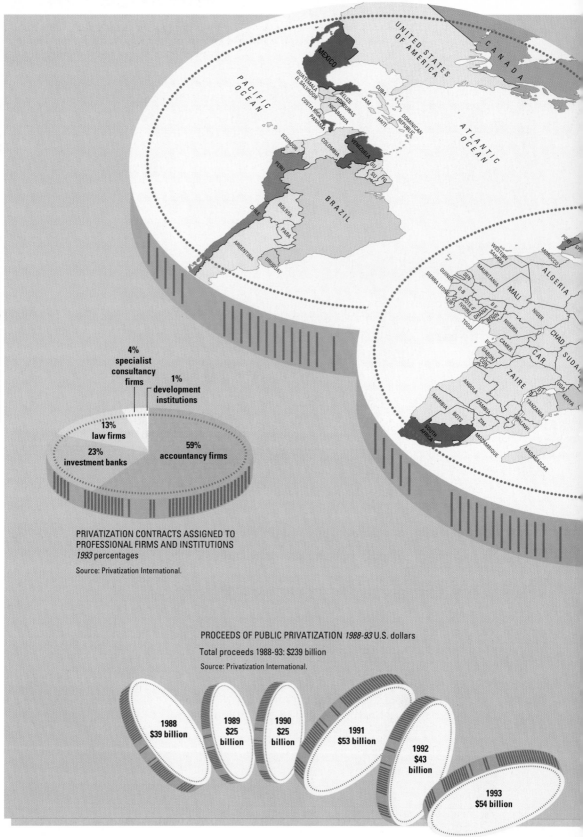

4%
specialist
consultancy
firms

1%
development
institutions

13%
law firms

23%
investment banks

59%
accountancy firms

PRIVATIZATION CONTRACTS ASSIGNED TO
PROFESSIONAL FIRMS AND INSTITUTIONS
1993 percentages

Source: Privatization International.

PROCEEDS OF PUBLIC PRIVATIZATION *1988-93* U.S. dollars

Total proceeds 1988-93: $239 billion

Source: Privatization International.

1988
$39 billion

1989
$25
billion

1990
$25
billion

1991
$53 billion

1992
$43
billion

1993
$54 billion

State assets worth a quarter of a trillion dollars (U.S. $250,000,000,000) went private publicly between 1988 and 1993. Another untold sum went private privately. The transfers continue.

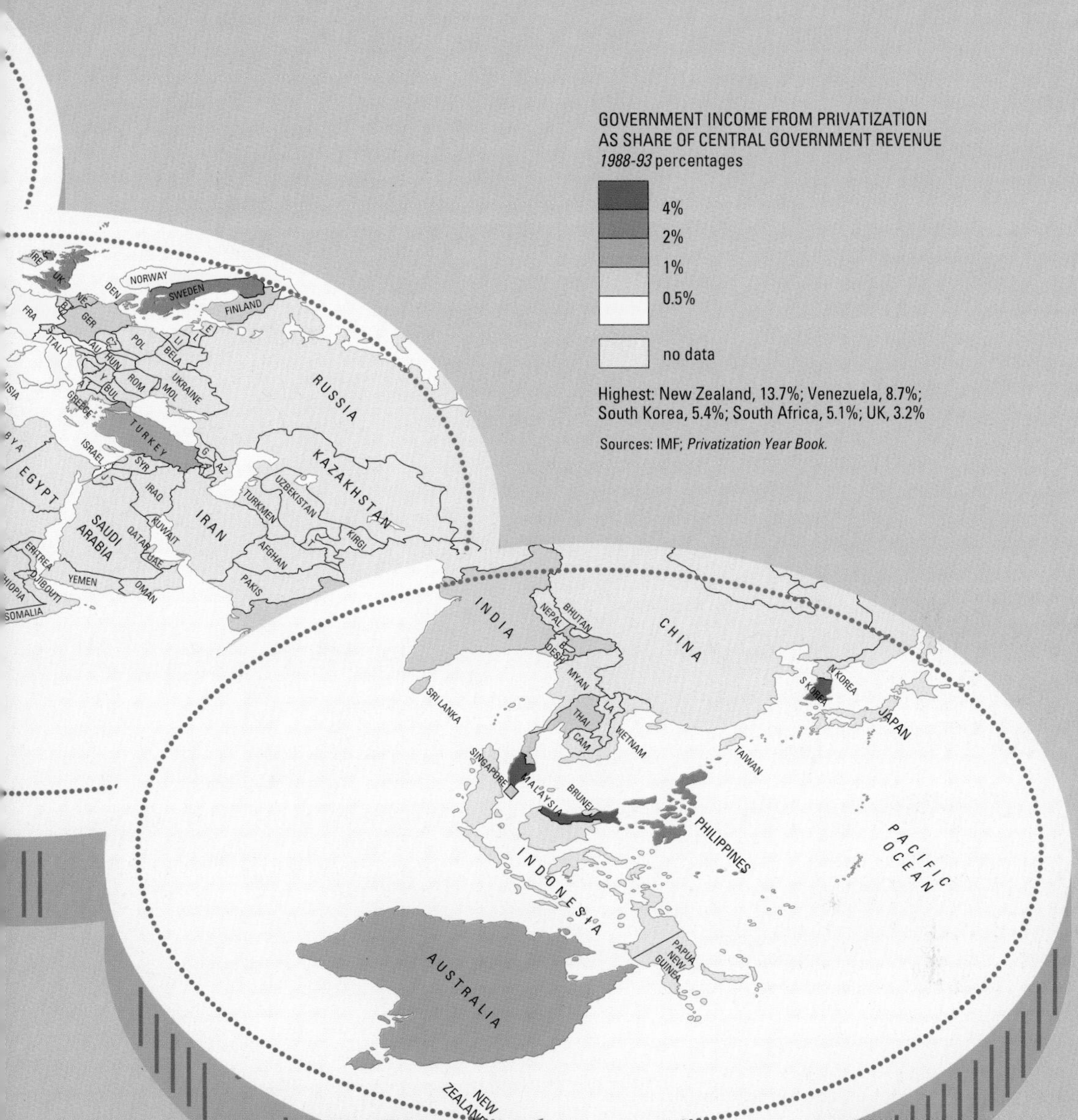

GOVERNMENT INCOME FROM PRIVATIZATION
AS SHARE OF CENTRAL GOVERNMENT REVENUE
1988-93 percentages

4%
2%
1%
0.5%

no data

Highest: New Zealand, 13.7%; Venezuela, 8.7%; South Korea, 5.4%; South Africa, 5.1%; UK, 3.2%

Sources: IMF; *Privatization Year Book.*

States are not equal citizens of the world of states. Embassies in states considered important are usually headed by a resident ambassador or equivalent. States considered less important are more likely to be headed by a Chargé d'Affaires or Head of Mission.

Diplomatic relations can be fairly notional: 6 of the 89 embassies in Teheran, Iran, were vacant at the end of 1992, as were 4 of the 28 in Kabul, Afghanistan. Many missions have to share ambassadors, or be serviced by proxies. Some are no more than post office boxes. Others proclaim their presence but give no clue as to their whereabouts – Namibia's mission in Yaoundé, Cameroon, for example, or the Marshall Islands' mission in Beijing, or Singapore's in Accra. Business in these places cannot be brisk.

Source: *Europa*.

21 states host between 0-10 ambassadors

79 states host between 11-50 ambassadors

46 states host between 51-100 ambassadors

9 states host between 101-150 ambassadors

GREENLAND
(Den)

ICELAND

CANADA

NORWAY SWEDEN

DENMARK

IRELAND UNITED
KINGDOM

NETH.
BELG. GERMANY POLAND
CZECH
FRANCE AUS HUNG
ITALY B-H YUG
AL

PORTUGAL SPAIN GREECE

UNITED STATES
OF AMERICA

MOROCCO

TUNISIA

ALGERIA LIBYA

WESTERN SAHARA

ATLANTIC
OCEAN

MEXICO

BAHAMAS

CUBA

DOMINICAN
REPUBLIC

BELIZE JAMAICA
HONDURAS HAITI
GUATEMALA
EL SALVADOR

ANTIGUA & BARBUDA

DOMINICA

CAPE VERDE

MAURITANIA

MALI
SENEGAL

NIGER

CHAD

NICARAGUA GRENADA

COSTA RICA
PANAMA

VENEZUELA GUYANA
SURINAME
FRENCH GUIANA (Fr)

TRINIDAD & TOBAGO

GAMBIA
GUINEA-BISSAU

GUINEA

BURKINA
FASO

BENIN

NIGERIA

CÔTE d'
IVOIRE GHANA

CAM

SIERRA LEONE

C A R

LIBERIA TOGO

PACIFIC
OCEAN

COLOMBIA

ECUADOR

EQUATORIAL
GUINEA
GABON
CONGO

ZAI

PERU

BRAZIL

ANGOLA

BOLIVIA

PARAGUAY

NAMIBIA

CHILE

URUGUAY

POLITICAL SYSTEMS *1993*

SOUTH
AFRICA

ARGENTINA

established multi-party,
formally democratic system

recently adopted multi-party, formally
democratic system; or in transition to one

one-party regime

military rule, in form or fact

monarchical and/or theocratic regime

disordered state: civil war or widespread
ethnic, clan or other conflict

dependent or occupied state or
assimilated territory

Sources: Keesing; *U.S. State Department Country Reports
on Human Rights Practices for 1993;* press reports.

Michael Kidron and Ronald Segal *The State of the World Atlas* 5th edition   Copyright © Myriad Editions Limited

78

Not all voters have equal voices. In some states, votes are simply invented or destroyed. In others, the gun, the throne or the supposed voice of God decides.

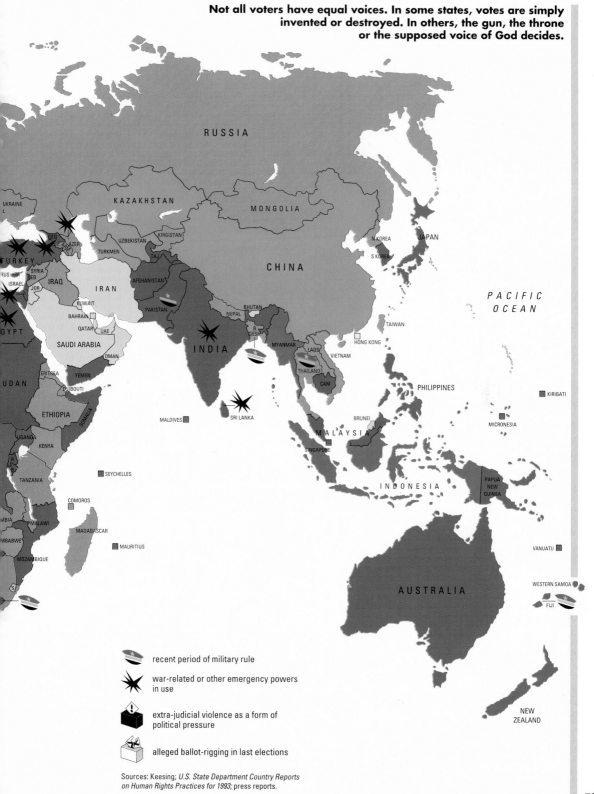

recent period of military rule

war-related or other emergency powers in use

extra-judicial violence as a form of political pressure

alleged ballot-rigging in last elections

Sources: Keesing; *U.S. State Department Country Reports on Human Rights Practices for 1993*; press reports.

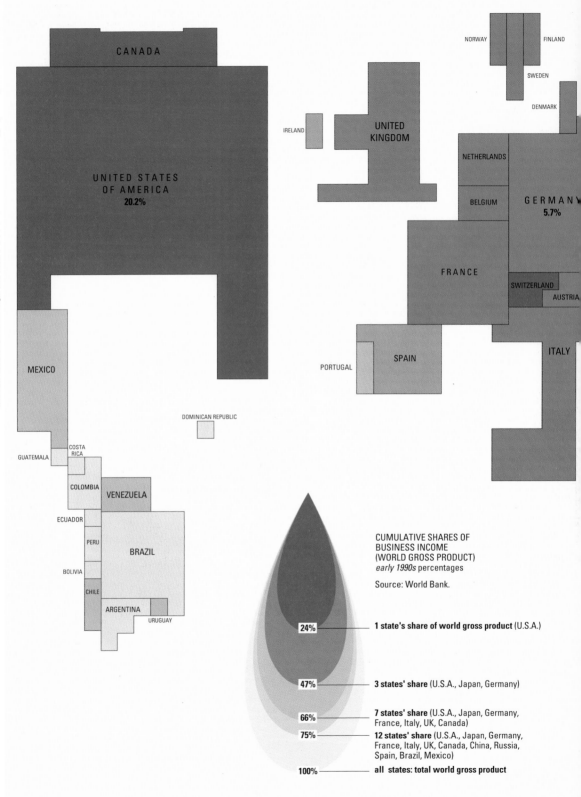

CANADA

UNITED STATES
OF AMERICA
20.2%

IRELAND

UNITED
KINGDOM

NORWAY

FINLAND

SWEDEN

DENMARK

NETHERLANDS

G E R M A N Y
5.7%

BELGIUM

FRANCE

SWITZERLAND

AUSTRIA

PORTUGAL

SPAIN

ITALY

MEXICO

DOMINICAN REPUBLIC

COSTA
RICA

GUATEMALA

COLOMBIA

VENEZUELA

ECUADOR

PERU

BRAZIL

BOLIVIA

CHILE

ARGENTINA

URUGUAY

CUMULATIVE SHARES OF
BUSINESS INCOME
(WORLD GROSS PRODUCT)
early 1990s percentages

Source: World Bank.

24% — **1 state's share of world gross product** (U.S.A.)

47% — **3 states' share** (U.S.A., Japan, Germany)

66% — **7 states' share** (U.S.A., Japan, Germany,
France, Italy, UK, Canada)

75% — **12 states' share** (U.S.A., Japan, Germany,
France, Italy, UK, Canada, China, Russia,
Spain, Brazil, Mexico)

100% — **all states: total world gross product**

80

Whichever way the slippery notion of national income is measured, a handful of states invariably emerges as engrossing the lion's share.

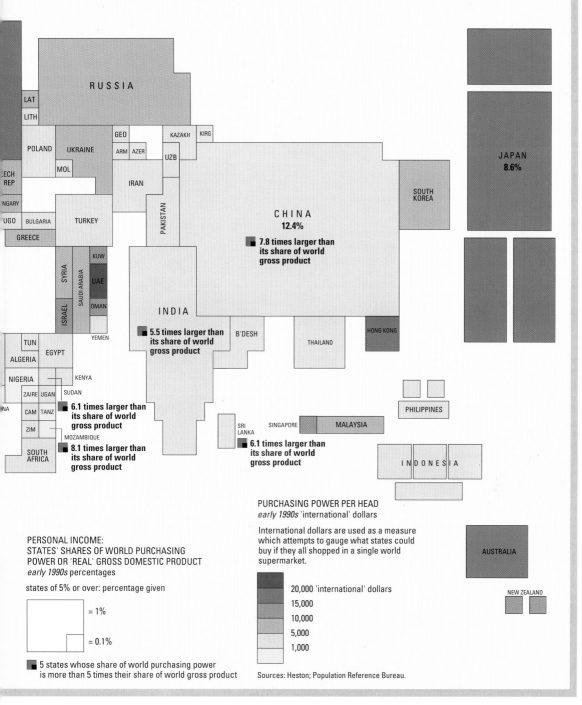

RUSSIA

LAT
LITH

POLAND | UKRAINE

GEO | KAZAKH | KIRG
ARM | AZER
UZB

.ECH REP

MOL

IRAN

NGARY

UGO | BULGARIA | TURKEY

GREECE

SYRIA | SAUDI ARABIA | KUW | UAE | OMAN

ISRAEL

YEMEN

PAKISTAN

KAZAKH | KIRG

SOUTH KOREA

JAPAN
8.6%

C H I N A
12.4%

■ **7.8 times larger than its share of world gross product**

I N D I A

■ **5.5 times larger than its share of world gross product**

B'DESH

THAILAND

HONG KONG

TUN
ALGERIA | EGYPT

NIGERIA | KENYA

ZAIRE | UGAN | SUDAN

NA | CAM | TANZ
ZIM

MOZAMBIQUE

SOUTH AFRICA

■ **6.1 times larger than its share of world gross product**

■ **8.1 times larger than its share of world gross product**

SRI LANKA | SINGAPORE | MALAYSIA

■ **6.1 times larger than its share of world gross product**

PHILIPPINES

I N D O N E S I A

AUSTRALIA

NEW ZEALAND

PURCHASING POWER PER HEAD
early 1990s 'international' dollars

International dollars are used as a measure which attempts to gauge what states could buy if they all shopped in a single world supermarket.

PERSONAL INCOME:
STATES' SHARES OF WORLD PURCHASING POWER OR 'REAL' GROSS DOMESTIC PRODUCT
early 1990s percentages

states of 5% or over: percentage given

☐ = 1%

▫ = 0.1%

■ 5 states whose share of world purchasing power is more than 5 times their share of world gross product

	20,000 'international' dollars
	15,000
	10,000
	5,000
	1,000

Sources: Heston; Population Reference Bureau.

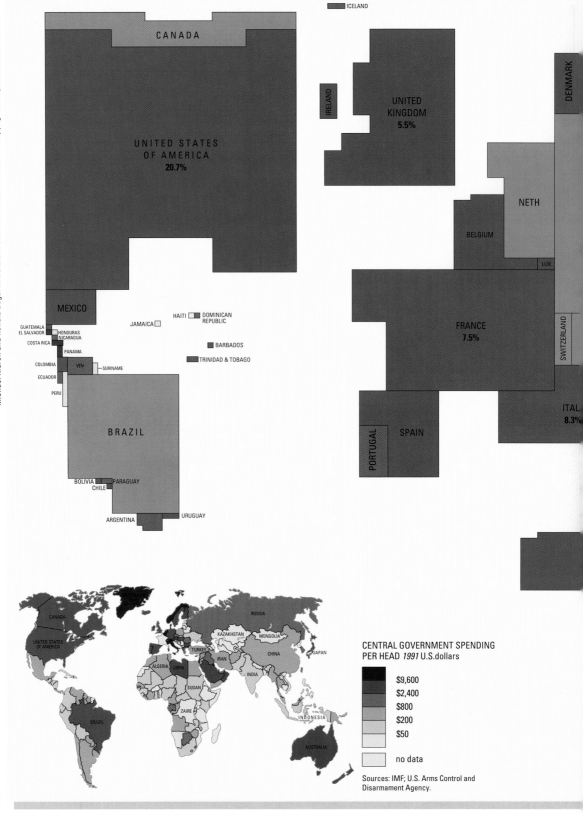

ICELAND

CANADA

IRELAND

UNITED
KINGDOM
5.5%

DENMARK

UNITED STATES
OF AMERICA
20.7%

NETH

BELGIUM

LUX

MEXICO

JAMAICA

HAITI

DOMINICAN
REPUBLIC

FRANCE
7.5%

SWITZERLAND

GUATEMALA
EL SALVADOR

HONDURAS
NICARAGUA

COSTA RICA

PANAMA

BARBADOS

TRINIDAD & TOBAGO

COLOMBIA

VEN

SURINAME

ECUADOR

PERU

ITAL
8.3%

BRAZIL

PORTUGAL

SPAIN

BOLIVIA

PARAGUAY

CHILE

ARGENTINA

URUGUAY

CANADA

RUSSIA

UNITED STATES
OF AMERICA

KAZAKHSTAN

MONGOLIA

TURKEY

CHINA

JAPAN

IRAN

ALGERIA

LIBYA

INDIA

SUDAN

ZAIRE

INDONESIA

BRAZIL

AUSTRALIA

CENTRAL GOVERNMENT SPENDING
PER HEAD *1991* U.S.dollars

$9,600
$2,400
$800
$200
$50

no data

Sources: IMF; U.S. Arms Control and
Disarmament Agency.

In 1991, national governments spent U.S. $7 trillion — 28 percent of world income. A quarter of their spending went on paying themselves.

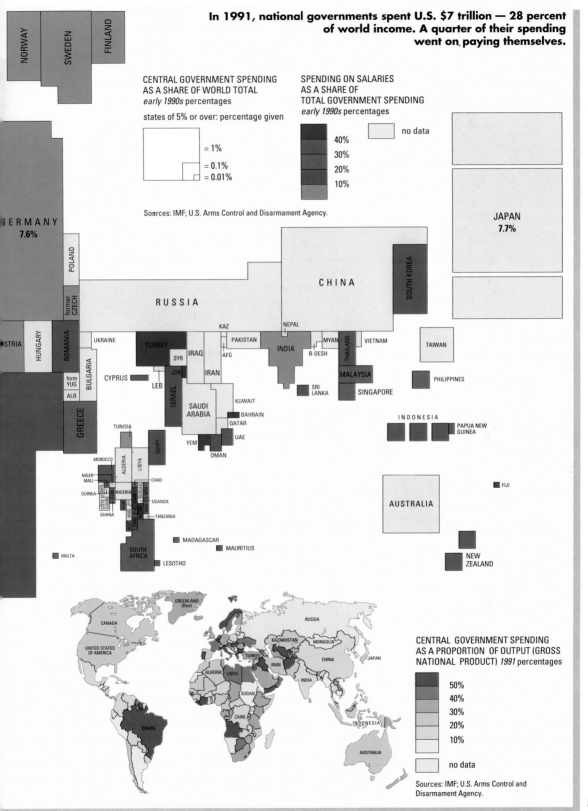

CENTRAL GOVERNMENT SPENDING
AS A SHARE OF WORLD TOTAL
early 1990s percentages

states of 5% or over: percentage given

= 1%
= 0.1%
= 0.01%

SPENDING ON SALARIES
AS A SHARE OF
TOTAL GOVERNMENT SPENDING
early 1990s percentages

no data

40%
30%
20%
10%

Sources: IMF; U.S. Arms Control and Disarmament Agency.

NORWAY
SWEDEN
FINLAND

GERMANY
7.6%

POLAND

former CZECH

RUSSIA

CHINA

SOUTH KOREA

JAPAN
7.7%

STRIA
HUNGARY
ROMANIA
UKRAINE

form YUG
BULGARIA
ALB
GREECE

TURKEY
SYR
IRAQ
JOR
IRAN
LEB
ISRAEL
CYPRUS

KAZ
PAKISTAN
AFG
NEPAL
INDIA
B-DESH
MYAN
VIETNAM
THAILAND
MALAYSIA
SINGAPORE

TAIWAN
PHILIPPINES

SAUDI
ARABIA
KUWAIT
BAHRAIN
QATAR
UAE
OMAN
YEM

SRI LANKA

INDONESIA
PAPUA NEW GUINEA

TUNISIA
MOROCCO
ALGERIA
LIBYA
EGYPT

NIGER
MALI
GUINEA
COTE D'IVOIRE
GHANA
NIGERIA
CHAD
SUDAN
UGANDA
TANZANIA

AUSTRALIA
FIJI

MADAGASCAR
MAURITIUS

MALTA
SOUTH AFRICA
LESOTHO

NEW ZEALAND

GREENLAND (Den)

CANADA

UNITED STATES
OF AMERICA

RUSSIA

KAZAKHSTAN
MONGOLIA
TURKEY
IRAN
CHINA
JAPAN

ALGERIA
LIBYA
INDIA

SUDAN

BRAZIL
ZAIRE

INDONESIA

AUSTRALIA

CENTRAL GOVERNMENT SPENDING
AS A PROPORTION OF OUTPUT (GROSS
NATIONAL PRODUCT) *1991* percentages

50%
40%
30%
20%
10%

no data

Sources: IMF; U.S. Arms Control and
Disarmament Agency.

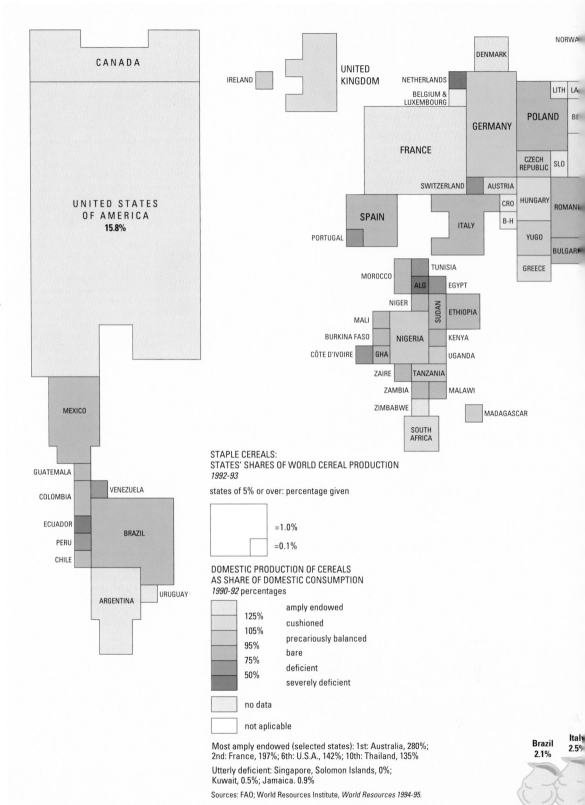

Michael Kidron and Ronald Segal *The State of the World Atlas* 5th edition   Copyright © Myriad Editions Limited

CANADA

UNITED STATES
OF AMERICA
15.8%

IRELAND

UNITED
KINGDOM

NETHERLANDS

BELGIUM &
LUXEMBOURG

FRANCE

GERMANY

DENMARK

NORWA

POLAND

LITH | LA

BE

CZECH
REPUBLIC

SLO

SWITZERLAND

AUSTRIA

CRO

HUNGARY

ROMANI

SPAIN

ITALY

B-H

YUGO

PORTUGAL

MOROCCO

TUNISIA

ALG

EGYPT

NIGER

MALI

BURKINA FASO

CÔTE D'IVOIRE

GHA

SUDAN

ETHIOPIA

NIGERIA

KENYA

UGANDA

ZAIRE

TANZANIA

ZAMBIA

MALAWI

ZIMBABWE

MADAGASCAR

SOUTH
AFRICA

GREECE

BULGARI

MEXICO

GUATEMALA

COLOMBIA

VENEZUELA

ECUADOR

PERU

CHILE

BRAZIL

ARGENTINA

URUGUAY

STAPLE CEREALS:
STATES' SHARES OF WORLD CEREAL PRODUCTION
1992-93

states of 5% or over: percentage given

☐ =1.0%

□ =0.1%

DOMESTIC PRODUCTION OF CEREALS
AS SHARE OF DOMESTIC CONSUMPTION
1990-92 percentages

125%   amply endowed
105%   cushioned
95%   precariously balanced
75%   bare
50%   deficient
        severely deficient

        no data

        not aplicable

Most amply endowed (selected states): 1st: Australia, 280%;
2nd: France, 197%; 6th: U.S.A., 142%; 10th: Thailand, 135%

Utterly deficient: Singapore, Solomon Islands, 0%;
Kuwait, 0.5%; Jamaica. 0.9%

Sources: FAO; World Resources Institute, *World Resources 1994-95*.

**Brazil
2.1%**

**Ital
2.5%**

A state that cannot guarantee its citizens adequate food is a state in crisis.

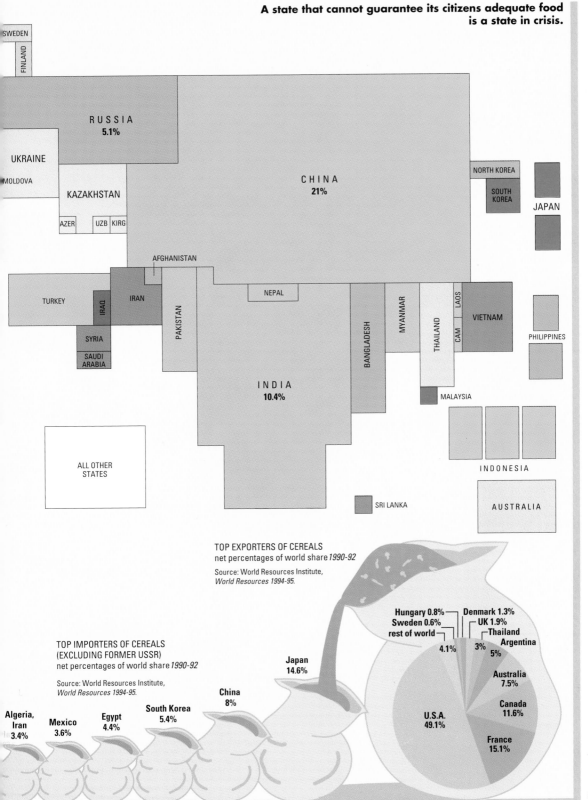

SWEDEN

FINLAND

RUSSIA
5.1%

UKRAINE

MOLDOVA

KAZAKHSTAN

AZER   UZB   KIRG

AFGHANISTAN

TURKEY   IRAQ   IRAN

SYRIA

SAUDI ARABIA

PAKISTAN

NEPAL

CHINA
21%

NORTH KOREA

SOUTH KOREA

JAPAN

INDIA
10.4%

BANGLADESH

MYANMAR

THAILAND

CAM   LAOS

VIETNAM

MALAYSIA

PHILIPPINES

SRI LANKA

INDONESIA

AUSTRALIA

ALL OTHER STATES

TOP EXPORTERS OF CEREALS
net percentages of world share *1990-92*

Source: World Resources Institute,
World Resources 1994-95.

TOP IMPORTERS OF CEREALS
(EXCLUDING FORMER USSR)
net percentages of world share *1990-92*

Source: World Resources Institute,
World Resources 1994-95.

Algeria,
Iran
3.4%

Mexico
3.6%

Egypt
4.4%

South Korea
5.4%

China
8%

Japan
14.6%

Hungary 0.8%   Denmark 1.3%
Sweden 0.6%   UK 1.9%
rest of world   Thailand
4.1%   3%   Argentina
                      5%

Australia
7.5%

Canada
11.6%

U.S.A.
49.1%

France
15.1%

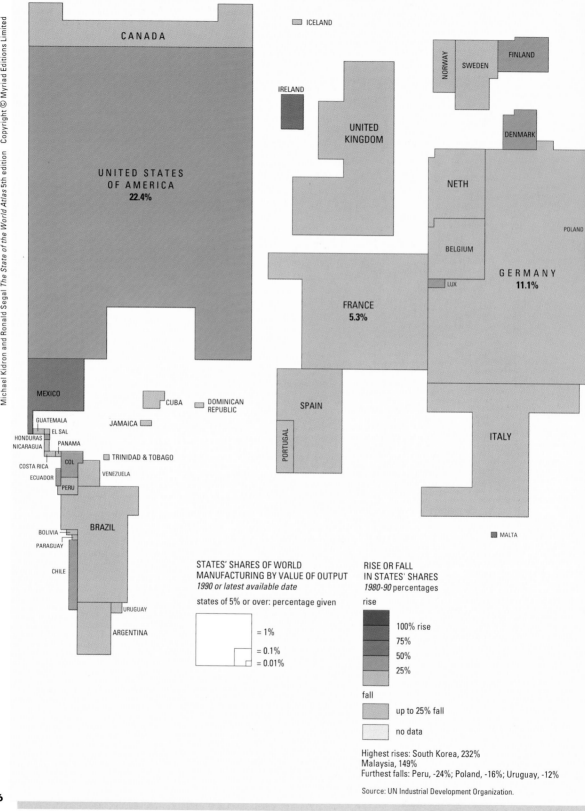

CANADA

ICELAND

NORWAY   SWEDEN   FINLAND

IRELAND

UNITED
KINGDOM

DENMARK

UNITED STATES
OF AMERICA
22.4%

NETH

POLAND

BELGIUM

GERMANY
11.1%

LUX

FRANCE
5.3%

MEXICO

CUBA

DOMINICAN
REPUBLIC

JAMAICA

GUATEMALA

EL SAL

SPAIN

HONDURAS

NICARAGUA

PANAMA

ITALY

COSTA RICA

COL

TRINIDAD & TOBAGO

PORTUGAL

ECUADOR

VENEZUELA

PERU

BOLIVIA

BRAZIL

PARAGUAY

MALTA

CHILE

URUGUAY

ARGENTINA

STATES' SHARES OF WORLD
MANUFACTURING BY VALUE OF OUTPUT
1990 or latest available date

states of 5% or over: percentage given

= 1%

= 0.1%

= 0.01%

RISE OR FALL
IN STATES' SHARES
1980-90 percentages

rise

100% rise

75%

50%

25%

fall

up to 25% fall

no data

Highest rises: South Korea, 232%
Malaysia, 149%
Furthest falls: Peru, -24%; Poland, -16%; Uruguay, -12%

Source: UN Industrial Development Organization.

The U.S.A. and Japan together account for over 40 percent of the world's total manufacturing output. Some states are rapidly growing industrial powers. Others are shrinking fast.

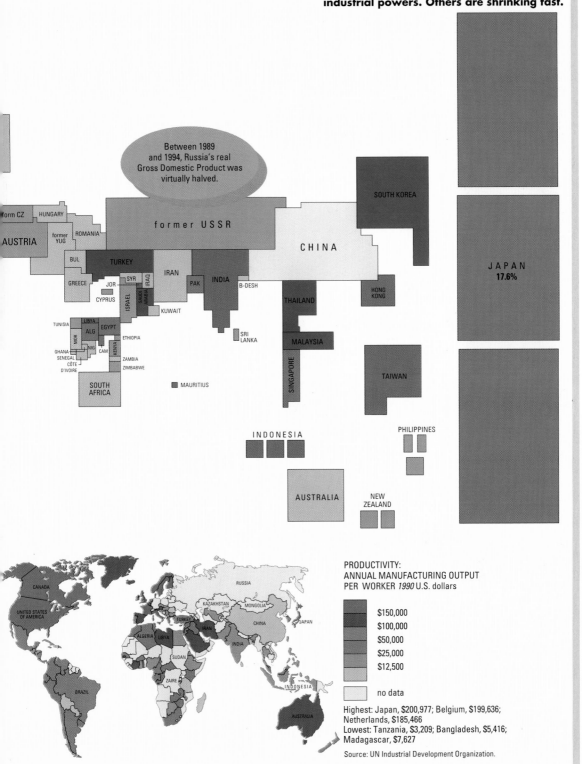

Between 1989 and 1994, Russia's real Gross Domestic Product was virtually halved.

former CZ

HUNGARY

form CZ

AUSTRIA

former YUG

ROMANIA

former USSR

BUL

TURKEY

GREECE

JOR

CYPRUS

SYR

IRAQ

ISRAEL

SAUDI ARABIA

KUWAIT

IRAN

PAK

INDIA

B-DESH

CHINA

SOUTH KOREA

HONG KONG

JAPAN
17.6%

THAILAND

TUNISIA

LIBYA

ALG

EGYPT

MOR

NIG

CAM

ETHIOPIA

SRI LANKA

MALAYSIA

GHANA

SENEGAL

CÔTE D'IVOIRE

KENYA

ZAMBIA

ZIMBABWE

SINGAPORE

TAIWAN

SOUTH AFRICA

MAURITIUS

INDONESIA

PHILIPPINES

AUSTRALIA

NEW ZEALAND

PRODUCTIVITY:
ANNUAL MANUFACTURING OUTPUT
PER WORKER *1990* U.S. dollars

- $150,000
- $100,000
- $50,000
- $25,000
- $12,500
- no data

CANADA

UNITED STATES OF AMERICA

BRAZIL

RUSSIA

KAZAKHSTAN

MONGOLIA

TURKEY

IRAN

CHINA

JAPAN

ALGERIA

LIBYA

SUDAN

INDIA

ZAIRE

INDONESIA

AUSTRALIA

Highest: Japan, $200,977; Belgium, $199,636; Netherlands, $185,466
Lowest: Tanzania, $3,209; Bangladesh, $5,416; Madagascar, $7,627

Source: UN Industrial Development Organization.

Michael Kidron and Ronald Segal *The State of the World Atlas* 5th edition   Copyright © Myriad Editions Limited

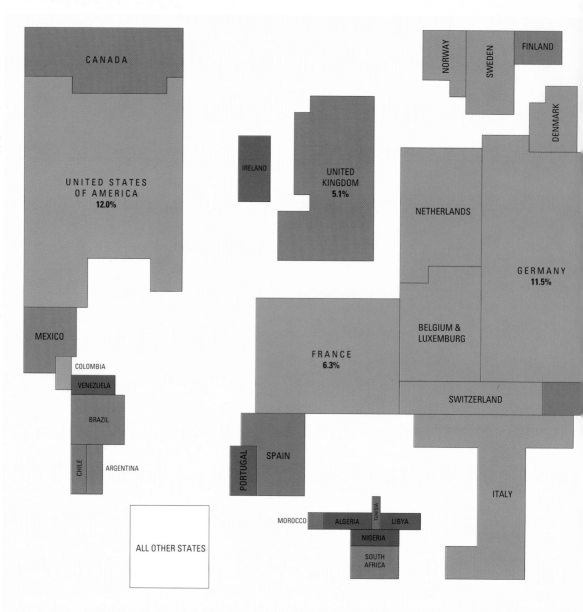

CANADA

UNITED STATES
OF AMERICA
12.0%

MEXICO

COLOMBIA

VENEZUELA

BRAZIL

CHILE   ARGENTINA

NORWAY   SWEDEN   FINLAND

DENMARK

IRELAND

UNITED
KINGDOM
5.1%

NETHERLANDS

GERMANY
11.5%

BELGIUM &
LUXEMBURG

FRANCE
6.3%

SWITZERLAND

PORTUGAL   SPAIN

ITALY

ALL OTHER STATES

MOROCCO   ALGERIA   TUNISIA   LIBYA

NIGERIA

SOUTH
AFRICA

STATES' SHARES OF WORLD TRADE
IN EXPORTED GOODS
1992 percentages

states of 5% or over: percentage given

=1.0%

=0.1%

RISE OR FALL
IN STATES' SHARES
1982 -92

rise

over two-thirds

one-third to two-thirds

up to one third

no change

fall

up to one third

one third to two-thirds

over two-thirds

not applicable

Highest rises: Hong Kong, 186%; Thailand, 135%; Portugal, 123%
Furthest falls: Kuwait, Saudi Arabia, 69%; Libya, 66%

Source: GATT.

Three states - Germany, Japan and the U.S.A. - account for virtually a third of all trade in exported goods.

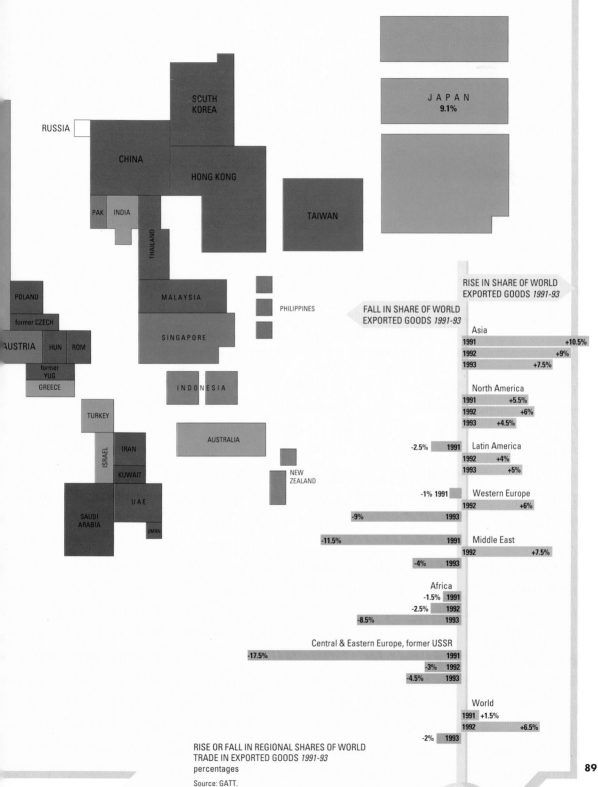

JAPAN
9.1%

SOUTH KOREA

RUSSIA

CHINA

HONG KONG

PAK | INDIA

TAIWAN

THAILAND

RISE IN SHARE OF WORLD
EXPORTED GOODS *1991-93*

POLAND

former CZECH

MALAYSIA

PHILIPPINES

FALL IN SHARE OF WORLD
EXPORTED GOODS *1991-93*

Asia
1991	+10.5%
1992	+9%
1993	+7.5%

AUSTRIA | HUN | ROM

SINGAPORE

former YUG

GREECE

INDONESIA

North America
1991	+5.5%
1992	+6%
1993	+4.5%

TURKEY

ISRAEL | IRAN

KUWAIT

AUSTRALIA

Latin America
-2.5%	1991
1992	+4%
1993	+5%

NEW ZEALAND

Western Europe
-1% 1991	
1992	+6%
-9%	1993

SAUDI ARABIA | UAE

OMAN

Middle East
-11.5%	1991
1992	+7.5%
-4%	1993

Africa
-1.5%	1991
-2.5%	1992
-8.5%	1993

Central & Eastern Europe, former USSR
-17.5%	1991
-3%	1992
-4.5%	1993

World
1991	+1.5%
1992	+6.5%
-2%	1993

RISE OR FALL IN REGIONAL SHARES OF WORLD
TRADE IN EXPORTED GOODS *1991-93*
percentages
Source: GATT.

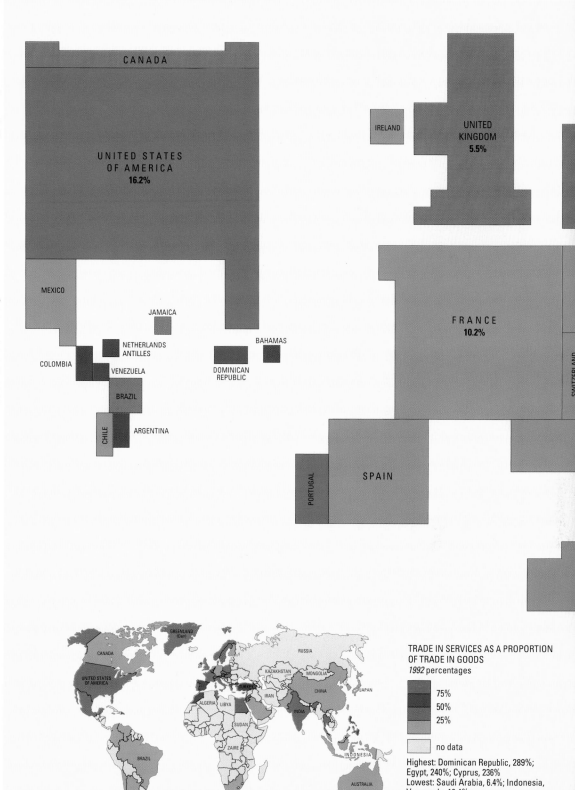

CANADA

UNITED STATES
OF AMERICA
16.2%

IRELAND

UNITED
KINGDOM
5.5%

MEXICO

JAMAICA

NETHERLANDS
ANTILLES

BAHAMAS

COLOMBIA

VENEZUELA

DOMINICAN
REPUBLIC

FRANCE
10.2%

SWITZERLAND

BRAZIL

CHILE

ARGENTINA

PORTUGAL

SPAIN

GREENLAND
(Den)

CANADA

RUSSIA

UNITED STATES
OF AMERICA

KAZAKHSTAN

MONGOLIA

JAPAN

CHINA

ALGERIA

LIBYA

TURKEY

IRAN

INDIA

SUDAN

ZAIRE

INDONESIA

BRAZIL

AUSTRALIA

TRADE IN SERVICES AS A PROPORTION
OF TRADE IN GOODS
1992 percentages

75%

50%

25%

no data

Highest: Dominican Republic, 289%;
Egypt, 240%; Cyprus, 236%
Lowest: Saudi Arabia, 6.4%; Indonesia,
Venezuela, 10.4%
Source: GATT.

Two states account for over a quarter, six states for nearly half, of all exports in commercial services.

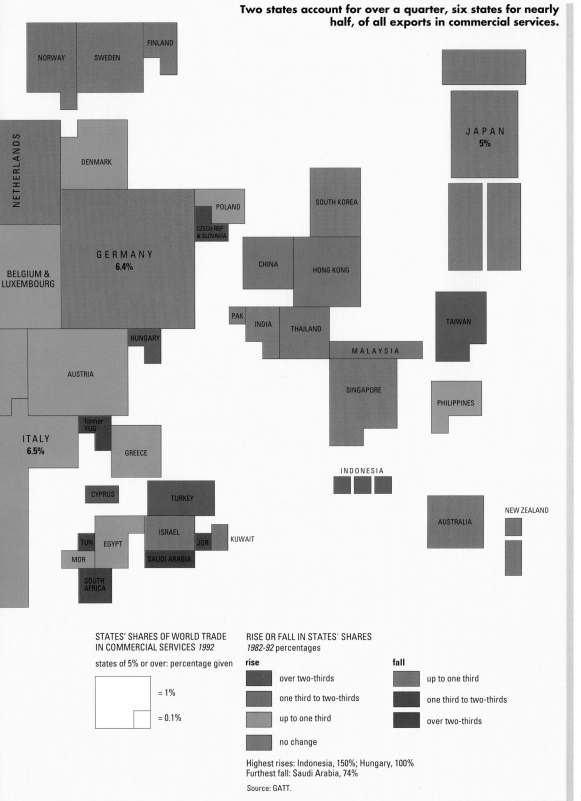

NORWAY

SWEDEN

FINLAND

NETHERLANDS

DENMARK

POLAND

CZECH REP & SLOVAKIA

SOUTH KOREA

JAPAN
5%

GERMANY
6.4%

BELGIUM & LUXEMBOURG

CHINA

HONG KONG

PAK

INDIA

THAILAND

TAIWAN

HUNGARY

AUSTRIA

MALAYSIA

SINGAPORE

PHILIPPINES

former YUG

ITALY
6.5%

GREECE

INDONESIA

CYPRUS

TURKEY

AUSTRALIA

NEW ZEALAND

TUN

EGYPT

ISRAEL

JOR

KUWAIT

MOR

SAUDI ARABIA

SOUTH AFRICA

STATES' SHARES OF WORLD TRADE
IN COMMERCIAL SERVICES *1992*

states of 5% or over: percentage given

= 1%

= 0.1%

RISE OR FALL IN STATES' SHARES
1982-92 percentages

rise

over two-thirds

one third to two-thirds

up to one third

no change

fall

up to one third

one third to two-thirds

over two-thirds

Highest rises: Indonesia, 150%; Hungary, 100%
Furthest fall: Saudi Arabia, 74%

Source: GATT.

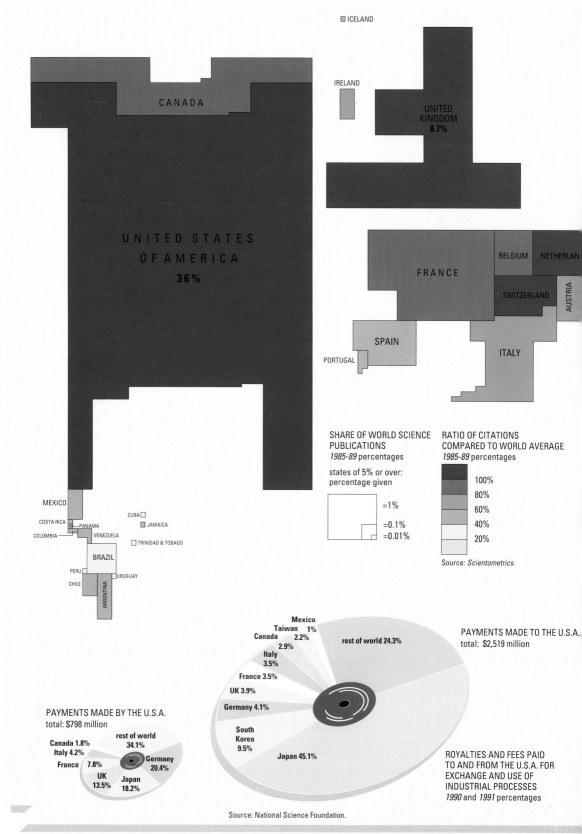

ICELAND

IRELAND

UNITED
KINGDOM
8.7%

CANADA

UNITED STATES
OF AMERICA
36%

BELGIUM   NETHERLAN

FRANCE

SWITZERLAND

AUSTRIA

SPAIN

PORTUGAL

ITALY

MEXICO

COSTA RICA

COLOMBIA

PANAMA

VENEZUELA

CUBA

JAMAICA

TRINIDAD & TOBAGO

BRAZIL

PERU

CHILE

URUGUAY

ARGENTINA

SHARE OF WORLD SCIENCE
PUBLICATIONS
1985-89 percentages

states of 5% or over:
percentage given

=1%

=0.1%

=0.01%

RATIO OF CITATIONS
COMPARED TO WORLD AVERAGE
1985-89 percentages

100%
80%
60%
40%
20%

Source: *Scientometrics.*

Mexico 1%
Taiwan
Canada 2.2%
2.9%
Italy
3.5%
France 3.5%
UK 3.9%
Germany 4.1%
South
Korea
9.5%
Japan 45.1%

rest of world 24.3%

PAYMENTS MADE TO THE U.S.A.
total: $2,519 million

PAYMENTS MADE BY THE U.S.A.
total: $798 million

Canada 1.8%
Italy 4.2%
France 7.8%
UK
13.5%
Japan
18.2%
Germany
20.4%
rest of world
34.1%

ROYALTIES AND FEES PAID
TO AND FROM THE U.S.A. FOR
EXCHANGE AND USE OF
INDUSTRIAL PROCESSES
1990 and *1991* percentages

92

Source: National Science Foundation.

The U.S.A. is the super science power — the biggest and most effective science producer and by far the biggest science market.

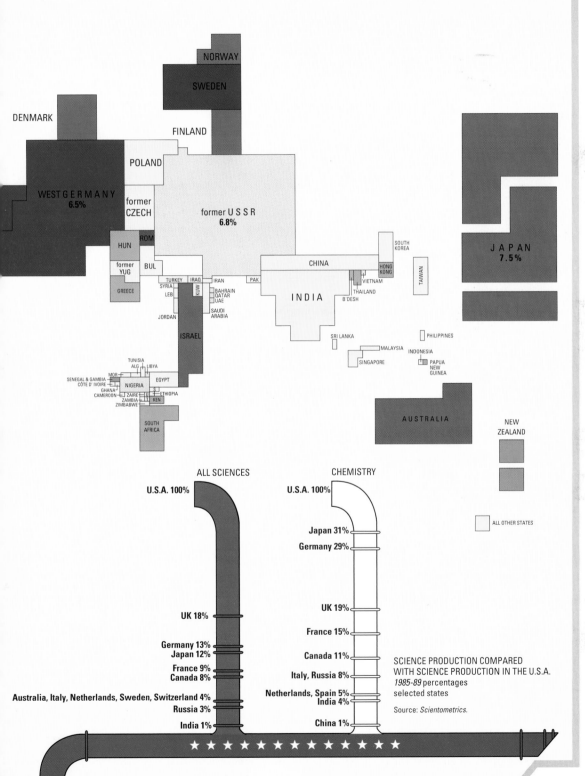

NORWAY

SWEDEN

DENMARK

FINLAND

POLAND

WEST GERMANY
6.5%

former
CZECH

former U S S R
6.8%

HUN

ROM

former
YUG

BUL

GREECE

SOUTH
KOREA

CHINA

HONG
KONG

TAIWAN

JAPAN
7.5%

TURKEY   IRAQ

SYRIA

LEB

KUW

IRAN

PAK

BAHRAIN
QATAR
UAE

VIETNAM

INDIA

THAILAND

B'DESH

JORDAN

SAUDI
ARABIA

ISRAEL

SRI LANKA

PHILIPPINES

MALAYSIA

INDONESIA

SINGAPORE

PAPUA
NEW
GUINEA

TUNISIA

ALG   LIBYA

MOR

SENEGAL & GAMBIA
CÔTE D' IVOIRE

EGYPT

GHANA
CAMEROON

NIGERIA

ZAIRE

ETHIOPIA

ZAMBIA   KEN
ZIMBABWE

SOUTH
AFRICA

AUSTRALIA

NEW
ZEALAND

ALL OTHER STATES

ALL SCIENCES

U.S.A. 100%

UK 18%

Germany 13%
Japan 12%
France 9%
Canada 8%

Australia, Italy, Netherlands, Sweden, Switzerland 4%

Russia 3%

India 1%

CHEMISTRY

U.S.A. 100%

Japan 31%

Germany 29%

UK 19%

France 15%

Canada 11%

Italy, Russia 8%

Netherlands, Spain 5%
India 4%

China 1%

SCIENCE PRODUCTION COMPARED
WITH SCIENCE PRODUCTION IN THE U.S.A.
1985-89 percentages
selected states

Source: *Scientometrics.*

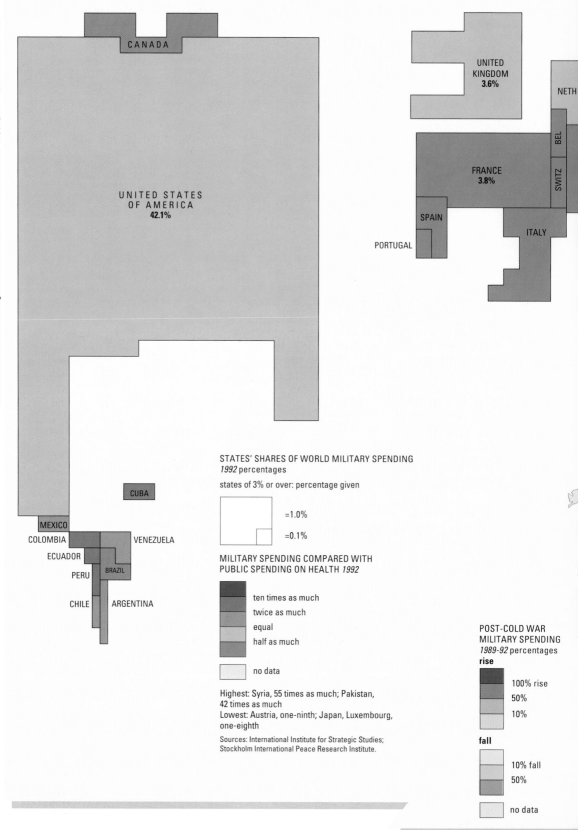

CANADA

UNITED STATES
OF AMERICA
42.1%

UNITED
KINGDOM
3.6%

NETH

FRANCE
3.8%

BEL

SWITZ

SPAIN

PORTUGAL

ITALY

CUBA

MEXICO

COLOMBIA

VENEZUELA

ECUADOR

BRAZIL

PERU

CHILE

ARGENTINA

STATES' SHARES OF WORLD MILITARY SPENDING
1992 percentages

states of 3% or over: percentage given

=1.0%

=0.1%

**MILITARY SPENDING COMPARED WITH
PUBLIC SPENDING ON HEALTH** *1992*

ten times as much

twice as much

equal

half as much

no data

Highest: Syria, 55 times as much; Pakistan,
42 times as much
Lowest: Austria, one-ninth; Japan, Luxembourg,
one-eighth

Sources: International Institute for Strategic Studies;
Stockholm International Peace Research Institute.

**POST-COLD WAR
MILITARY SPENDING**
1989-92 percentages
rise

100% rise

50%

10%

fall

10% fall

50%

no data

The military take U.S. $600 billion a year, a million dollars
a minute and more than three-quarters as much
as public spending on health.

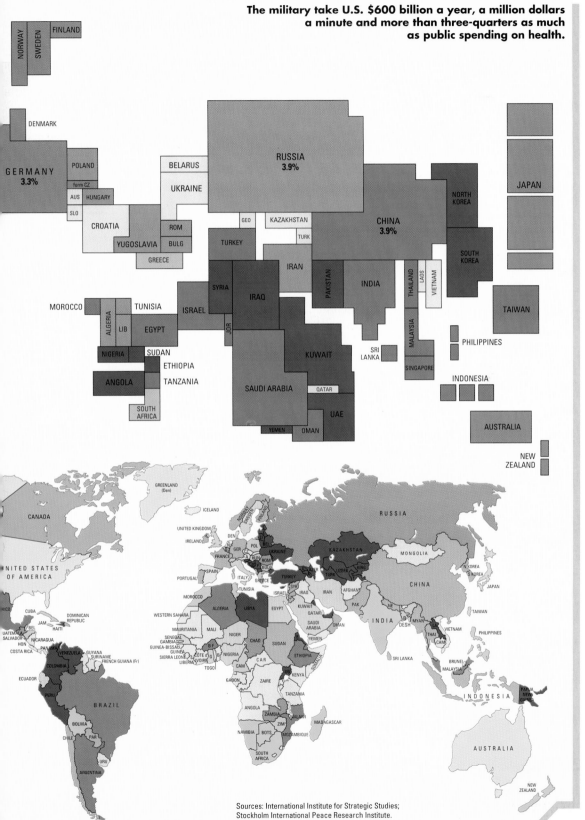

Sources: International Institute for Strategic Studies;
Stockholm International Peace Research Institute.

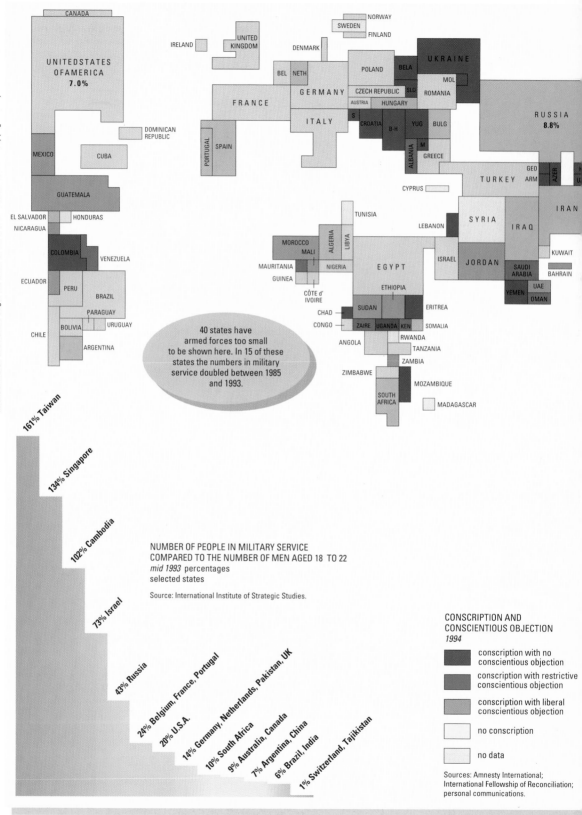

CANADA

UNITED STATES
OF AMERICA
7.0%

IRELAND

UNITED
KINGDOM

NORWAY

SWEDEN

FINLAND

DENMARK

BEL NETH

POLAND

BELA

UKRAINE

MOL

GERMANY

CZECH REPUBLIC

SLO

ROMANIA

AUSTRIA   HUNGARY

FRANCE

ITALY

S

CROATIA

B-H

YUG

BULG

RUSSIA
8.8%

PORTUGAL

SPAIN

ALBANIA

M

GREECE

GEO

ARM

AZER

U.

TURKEY

IRAN

CYPRUS

DOMINICAN
REPUBLIC

MEXICO

CUBA

GUATEMALA

EL SALVADOR   HONDURAS

NICARAGUA

COLOMBIA

VENEZUELA

ECUADOR

PERU

BRAZIL

PARAGUAY

BOLIVIA   URUGUAY

CHILE

ARGENTINA

TUNISIA

LEBANON

SYRIA

IRAQ

KUWAIT

MOROCCO

MALI

ALGERIA

LIBYA

ISRAEL

JORDAN

SAUDI
ARABIA

BAHRAIN

MAURITANIA

NIGERIA

EGYPT

YEMEN

UAE

OMAN

GUINEA

CÔTE d'
IVOIRE

ETHIOPIA

CHAD

SUDAN

ERITREA

CONGO

ZAIRE   UGANDA   KEN

SOMALIA

RWANDA

ANGOLA

TANZANIA

ZAMBIA

ZIMBABWE

MOZAMBIQUE

SOUTH
AFRICA

MADAGASCAR

40 states have
armed forces too small
to be shown here. In 15 of these
states the numbers in military
service doubled between 1985
and 1993.

161% Taiwan

134% Singapore

102% Cambodia

73% Israel

43% Russia

24% Belgium, France, Portugal

20% U.S.A.

14% Germany, Netherlands, Pakistan, UK

10% South Africa

9% Australia, Canada

7% Argentina, China

6% Brazil, India

1% Switzerland, Tajikistan

NUMBER OF PEOPLE IN MILITARY SERVICE
COMPARED TO THE NUMBER OF MEN AGED 18 TO 22
mid 1993 percentages
selected states

Source: International Institute of Strategic Studies.

CONSCRIPTION AND
CONSCIENTIOUS OBJECTION
1994

conscription with no
conscientious objection

conscription with restrictive
conscientious objection

conscription with liberal
conscientious objection

no conscription

no data

Sources: Amnesty International;
International Fellowship of Reconciliation;
personal communications.

30 million people are under arms fulltime in the world - more than 10 percent of the number of men aged 18 to 22.

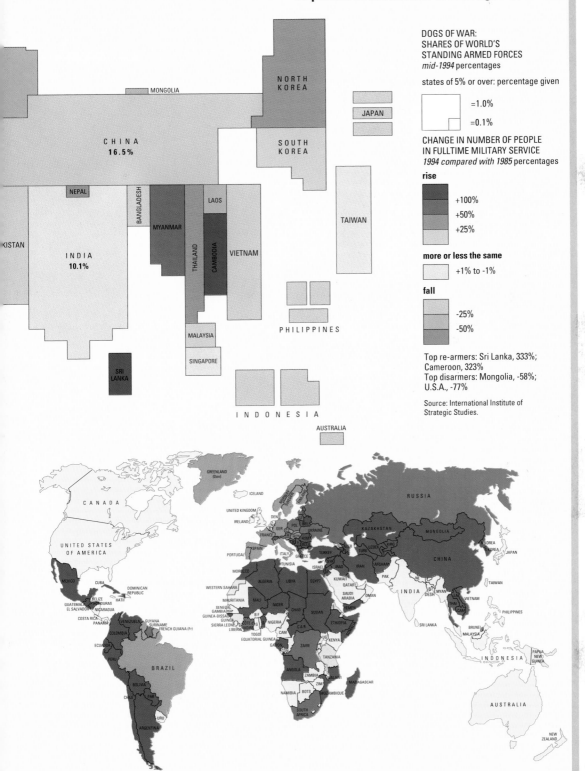

DOGS OF WAR:
SHARES OF WORLD'S
STANDING ARMED FORCES
mid-1994 percentages

states of 5% or over: percentage given

	=1.0%
	=0.1%

CHANGE IN NUMBER OF PEOPLE
IN FULLTIME MILITARY SERVICE
1994 compared with 1985 percentages

rise

+100%
+50%
+25%

more or less the same

+1% to -1%

fall

-25%
-50%

Top re-armers: Sri Lanka, 333%;
Cameroon, 323%
Top disarmers: Mongolia, -58%;
U.S.A., -77%

Source: International Institute of
Strategic Studies.

MONGOLIA

NORTH
KOREA

JAPAN

CHINA
16.5%

SOUTH
KOREA

NEPAL

BANGLADESH

LAOS

MYANMAR

TAIWAN

KISTAN

THAILAND

CAMBODIA

VIETNAM

INDIA
10.1%

MALAYSIA

SINGAPORE

PHILIPPINES

SRI
LANKA

INDONESIA

AUSTRALIA

Michael Kidron and Ronald Segal *The State of the World Atlas* 5th edition   Copyright © Myriad Editions Limited

CANADA

GREENLAND
(Den)

ICELAND

12 states   38 states

UNITED STATES
OF AMERICA

MEXICO

CUBA

DOMINICAN
REPUBLIC

JAMAICA
BELIZE
HONDURAS
GUATEMALA
EL SALVADOR
NICARAGUA

HAITI

COSTA RICA
PANAMA

ATLANTIC
OCEAN

VENEZUELA

TRINIDAD & TOBAGO

GUYANA
SURINAME
FRENCH GUIANA (Fr)

COLOMBIA

ECUADOR

PACIFIC
OCEAN

BRAZIL

PERU

BOLIVIA

PARAGUAY

CHILE

URUGUAY

ARGENTINA

NORWAY
SWEDEN

UNITED
KINGDOM
IRELAND

DENMARK

NETH
BEL
GERMANY
CZECH
AUS   HUN
S
B-H   YUG
FRANCE   ITALY   R
AL   B

POLAND
SLO

PORTUGAL   SPAIN

MOROCCO

TUNISIA

WESTERN SAHARA

ALGERIA

LIBYA

MAURITANIA

MALI

NIGER

CHAD

SENEGAL
GAMBIA
GUINEA-BISSAU
GUINEA
SIERRA LEONE
LIBERIA

BURKINA
FASO
CÔTE d'
IVOIRE
GHANA
TOGO
BENIN

NIGERIA

CAMEROON

EQUATORIAL GUINEA

CAR

GABON
CONGO

ZAI

ANGOLA

NAMIBIA

SOUTH
AFRIC

LICENSED KILLING:
LEGAL STATUS OF CAPITAL PUNISHMENT
early 1990s

[ ] abolished for all crimes

[ ] abolished for ordinary crimes but retained
for crimes against the state and its agents,
or for crimes committed in extraordinary
times, such as war

[ ] retained for ordinary crimes but
no executions since 1980 or independence

[ ] retained and used for ordinary crimes

[ ] federal system: status varies

UNLICENSED KILLING

extra-judicial killing and/or officially-inspired
'disappearances' reported to Amnesty International

Sources: Amnesty International; U.S. State Department.

98

**States deal harshly with their citizens.
Many terrorize them.**

RUSSIA

UKRAINE

KAZAKHSTAN

MONGOLIA

JAPAN

N.KOREA

S.KOREA

TURKEY

GEO

AZER

UZBEKISTAN

KIRGISTAN

TURKMEN

TAJ.

CHINA

CYPRUS

SYRIA

LEB

ISRAEL

JOR

IRAQ

IRAN

AFGHANISTAN

KUWAIT

BAHRAIN

PAKISTAN

BHUTAN

NEPAL

PACIFIC
OCEAN

QATAR

UAE

EGYPT

SAUDI ARABIA

OMAN

INDIA

B'DESH

MYANMAR

LAOS

VIETNAM

TAIWAN

HONG KONG

ERITREA

YEMEN

THAILAND

CAM

PHILIPPINES

SUDAN

DJIBOUTI

ETHIOPIA

SOMALIA

MALDIVES

SRI LANKA

BRUNEI

UGANDA

KENYA

MALAYSIA

SINGAPORE

TANZANIA

COMOROS

INDONESIA

PAPUA
NEW
GUINEA

MBIA

MALAWI

MADAGASCAR

MAURITIUS

MBABWE

MOZAMBIQUE

AUSTRALIA

FIJI

COOK ISLANDS

NEW
ZEALAND

CANADA

RUSSIA

U.S.A.

KAZAKHSTAN

MONGOLIA

TURKEY

JAPAN

IRAN

CHINA

ALGERIA

LIBYA

INDIA

SUDAN

BRAZIL

ZAIRE

INDONESIA

AUSTRALIA

**TORTURE BY STATE EMPLOYEES
(POLICE, THE MILITARY OR PRISON OFFICERS)**
1991-93

one or more case of torture
reported to Amnesty International

other states or no data

Source: Amnesty International.

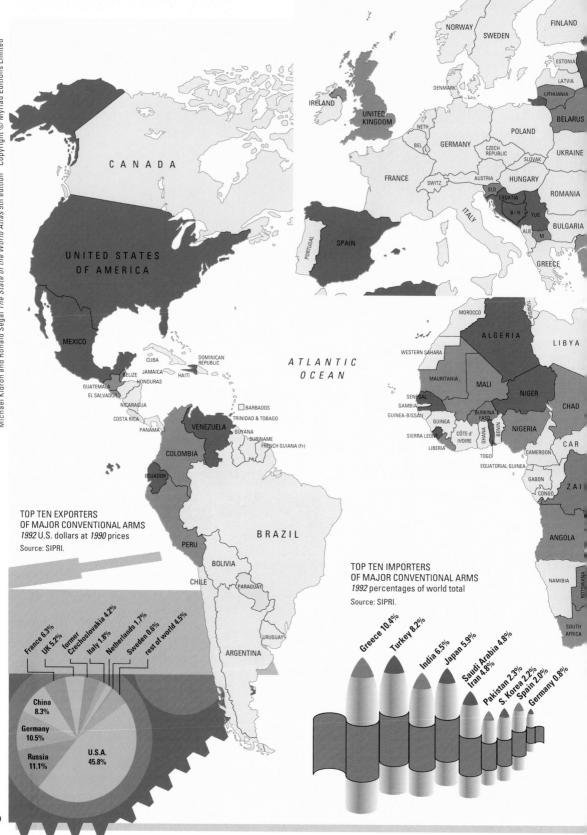

Michael Kidron and Ronald Segal *The State of the World Atlas* 5th edition   Copyright © Myriad Editions Limited

NORWAY
SWEDEN
FINLAND
DENMARK
ESTONIA
LATVIA
LITHUANIA
IRELAND
UNITED KINGDOM
NETH
BEL
GERMANY
POLAND
BELARUS
CZECH REPUBLIC
UKRAINE
FRANCE
SWITZ
AUSTRIA
SLOVAK
HUNGARY
SLO
CROATIA
ROMANIA
ITALY
B - H
YUG
BULGARIA
PORTUGAL
SPAIN
ALB
M
GREECE

CANADA

UNITED STATES OF AMERICA

MEXICO

CUBA
JAMAICA
BELIZE
HONDURAS
GUATEMALA
EL SALVADOR
NICARAGUA
COSTA RICA
PANAMA
DOMINICAN REPUBLIC
HAITI
BARBADOS
TRINIDAD & TOBAGO

ATLANTIC OCEAN

VENEZUELA
GUYANA
SURINAME
FRENCH GUIANA (Fr)
COLOMBIA
ECUADOR

MOROCCO
TUNISIA
ALGERIA
LIBYA
WESTERN SAHARA
MAURITANIA
MALI
NIGER
CHAD
SENEGAL
GAMBIA
GUINEA-BISSAU
GUINEA
BURKINA FASO
BENIN
NIGERIA
SIERRA LEONE
CÔTE d'IVOIRE
GHANA
CAMEROON
CAR
LIBERIA
TOGO
EQUATORIAL GUINEA
GABON
CONGO
ZAI

PERU
BRAZIL
BOLIVIA
CHILE
PARAGUAY
URUGUAY
ARGENTINA

ANGOLA
NAMIBIA
BOTSWANA
SOUTH AFRICA

TOP TEN EXPORTERS OF MAJOR CONVENTIONAL ARMS
1992 U.S. dollars at *1990* prices
Source: SIPRI.

France 6.3%
UK 5.2%
former Czechoslovakia 4.2%
Italy 1.8%
Netherlands 1.7%
Sweden 0.8%
rest of world 4.5%

China 8.3%
Germany 10.5%
Russia 11.1%
U.S.A. 45.8%

TOP TEN IMPORTERS OF MAJOR CONVENTIONAL ARMS
1992 percentages of world total
Source: SIPRI.

Greece 10.4%
Turkey 8.2%
India 6.5%
Japan 5.9%
Saudi Arabia 4.8%
Iran 4.8%
Pakistan 2.3%
S. Korea 2.2%
Spain 2.0%
Germany 0.8%

During 1994 more than 35 states were waging war inside or outside their territories. Altogether more than 50 states were at war at some time between 1990 and 1994.

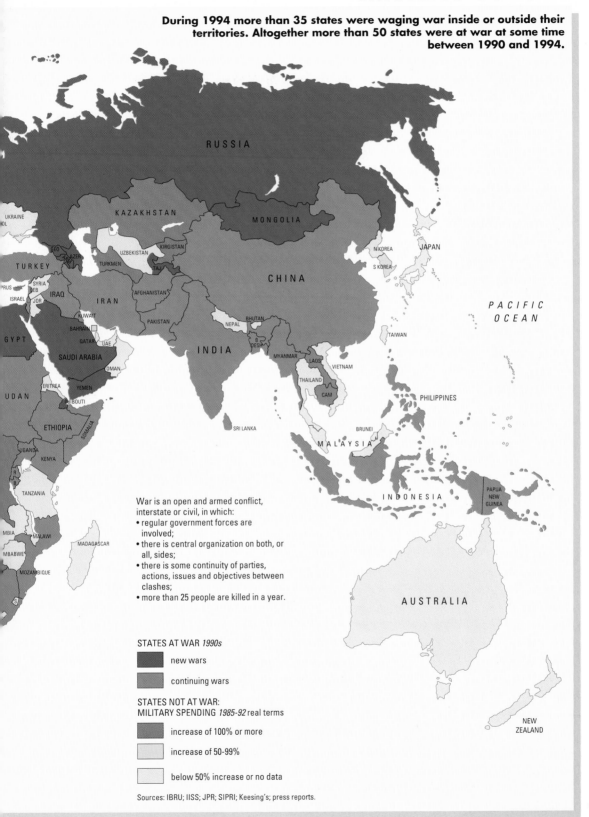

War is an open and armed conflict, interstate or civil, in which:
- regular government forces are involved;
- there is central organization on both, or all, sides;
- there is some continuity of parties, actions, issues and objectives between clashes;
- more than 25 people are killed in a year.

STATES AT WAR *1990s*

- new wars
- continuing wars

STATES NOT AT WAR:
MILITARY SPENDING *1985-92* real terms

- increase of 100% or more
- increase of 50-99%
- below 50% increase or no data

Sources: IBRU; IISS; JPR; SIPRI; Keesing's; press reports.

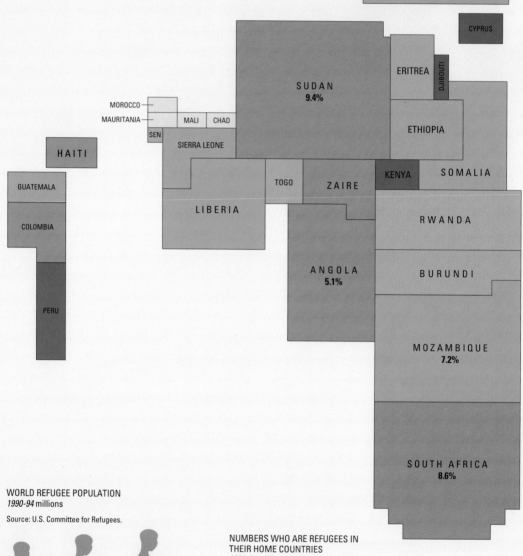

former
YUGOSLAVIA
6.4%

CYPRUS

ERITREA

DJIBOUTI

SUDAN
9.4%

MOROCCO —

MAURITANIA —

MALI   CHAD

SEN

ETHIOPIA

SIERRA LEONE

HAITI

GUATEMALA

COLOMBIA

PERU

TOGO

ZAIRE

KENYA

SOMALIA

LIBERIA

RWANDA

ANGOLA
5.1%

BURUNDI

MOZAMBIQUE
7.2%

SOUTH AFRICA
8.6%

WORLD REFUGEE POPULATION
1990-94 millions

Source: U.S. Committee for Refugees.

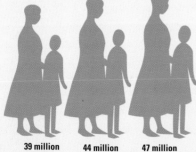

39 million
1990

44 million
1992

47 million
1994

NUMBERS WHO ARE REFUGEES IN
THEIR HOME COUNTRIES
1990-94 millions

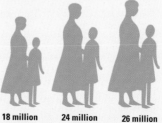

18 million
1990

24 million
1992

26 million
1994

At least 47 million people have abandoned their homes in fear for their lives and livelihoods. More than half are refugees in their own countries.

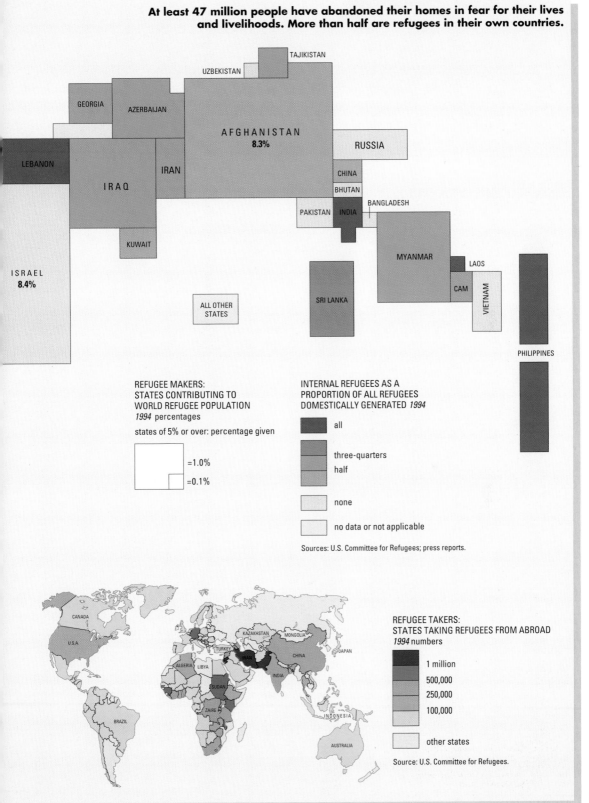

TAJIKISTAN

UZBEKISTAN

GEORGIA

AZERBAIJAN

AFGHANISTAN
8.3%

RUSSIA

LEBANON

IRAN

CHINA

BHUTAN

BANGLADESH

IRAQ

PAKISTAN

INDIA

KUWAIT

MYANMAR

LAOS

ISRAEL
8.4%

CAM

VIETNAM

ALL OTHER
STATES

SRI LANKA

PHILIPPINES

REFUGEE MAKERS:
STATES CONTRIBUTING TO
WORLD REFUGEE POPULATION
1994 percentages

states of 5% or over: percentage given

=1.0%

=0.1%

INTERNAL REFUGEES AS A
PROPORTION OF ALL REFUGEES
DOMESTICALLY GENERATED *1994*

all

three-quarters

half

no data or not applicable

Sources: U.S. Committee for Refugees; press reports.

CANADA

U.S.A.

KAZAKHSTAN

MONGOLIA

TURKEY

CHINA

JAPAN

ALGERIA

LIBYA

IRAN

INDIA

SUDAN

ZAIRE

INDONESIA

BRAZIL

AUSTRALIA

REFUGEE TAKERS:
STATES TAKING REFUGEES FROM ABROAD
1994 numbers

1 million

500,000

250,000

100,000

other states

Source: U.S. Committee for Refugees.

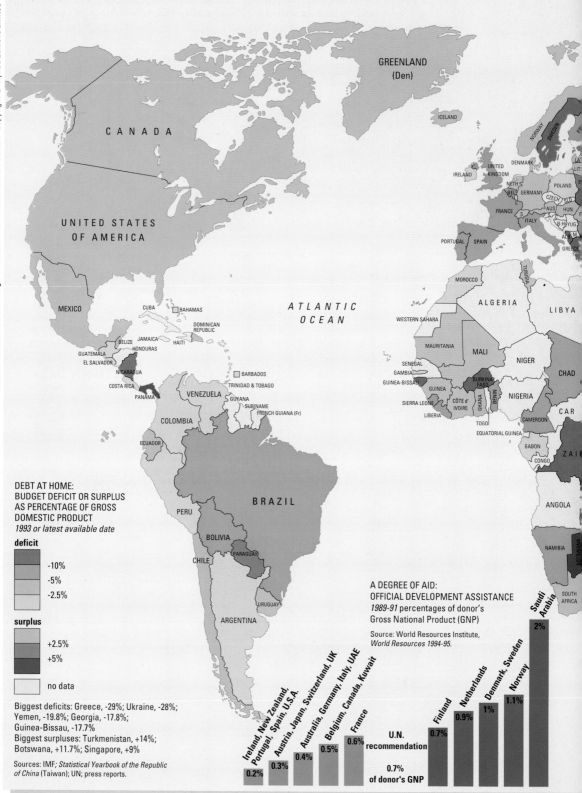

Michael Kidron and Ronald Segal *The State of the World Atlas* 5th edition   Copyright © Myriad Editions Limited

GREENLAND
(Den)

ICELAND

CANADA

UNITED STATES
OF AMERICA

MEXICO

CUBA
BAHAMAS
DOMINICAN
REPUBLIC
BELIZE
JAMAICA
HAITI
GUATEMALA
HONDURAS
EL SALVADOR
NICARAGUA
BARBADOS
COSTA RICA
TRINIDAD & TOBAGO
PANAMA
VENEZUELA
GUYANA
SURINAME
FRENCH GUIANA (Fr)
COLOMBIA
ECUADOR

PERU

BRAZIL

BOLIVIA
PARAGUAY
CHILE

URUGUAY
ARGENTINA

ATLANTIC
OCEAN

IRELAND
UNITED
KINGDOM
NETH
BEL
GERMANY
FRANCE
AUS
ITALY
PORTUGAL
SPAIN

DENMARK
POLAND
CZECH
SLU
HUN
B-H YUG
ALB M
GREECE

LA
LIT
FI

NORWAY
SWEDEN

MOROCCO
TUNISIA
ALGERIA
LIBYA

WESTERN SAHARA

MAURITANIA
MALI
NIGER
CHAD

SENEGAL
GAMBIA
GUINEA-BISSAU
GUINEA
BURKINA
FASO
SIERRA LEONE
CÔTE d'
IVOIRE
GHANA
BENIN
NIGERIA
LIBERIA
TOGO
CAMEROON
CAR
EQUATORIAL GUINEA
GABON
CONGO
ZAI

ANGOLA

NAMIBIA
BOTSWANA

SOUTH
AFRICA

Saudi
Arabia

DEBT AT HOME:
BUDGET DEFICIT OR SURPLUS
AS PERCENTAGE OF GROSS
DOMESTIC PRODUCT
1993 or latest available date

deficit

-10%

-5%

-2.5%

surplus

+2.5%

+5%

no data

Biggest deficits: Greece, -29%; Ukraine, -28%;
Yemen, -19.8%; Georgia, -17.8%;
Guinea-Bissau, -17.7%
Biggest surpluses: Turkmenistan, +14%;
Botswana, +11.7%; Singapore, +9%

Sources: IMF; *Statistical Yearbook of the Republic
of China* (Taiwan); UN; press reports.

A DEGREE OF AID:
OFFICIAL DEVELOPMENT ASSISTANCE
1989-91 percentages of donor's
Gross National Product (GNP)

Source: World Resources Institute,
World Resources 1994-95.

Ireland, New Zealand, Portugal, Spain, U.S.A. **0.2%**

Austria, Japan, Switzerland UK **0.3%**

Australia, Germany, Italy, UAE **0.4%**

Belgium, Canada, Kuwait **0.5%**

France **0.6%**

U.N.
recommendation

0.7%
of donor's GNP

Finland **0.7%**

Netherlands **0.9%**

Denmark, Sweden **1%**

Norway **1.1%**

Saudi Arabia **2%**

Governments frequently spend more than they collect in taxes. To finance deficits, they accumulate debt. To repay debt or meet interest payments, poor states sacrifice their export income.

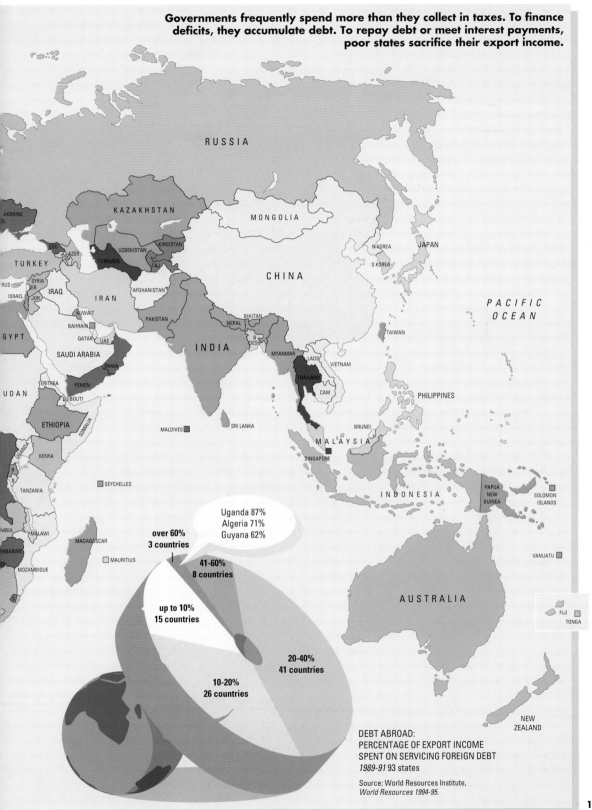

Uganda 87%
Algeria 71%
Guyana 62%

over 60%
3 countries

41-60%
8 countries

up to 10%
15 countries

20-40%
41 countries

10-20%
26 countries

DEBT ABROAD:
PERCENTAGE OF EXPORT INCOME
SPENT ON SERVICING FOREIGN DEBT
1989-91 93 states

Source: World Resources Institute,
World Resources 1994-95.

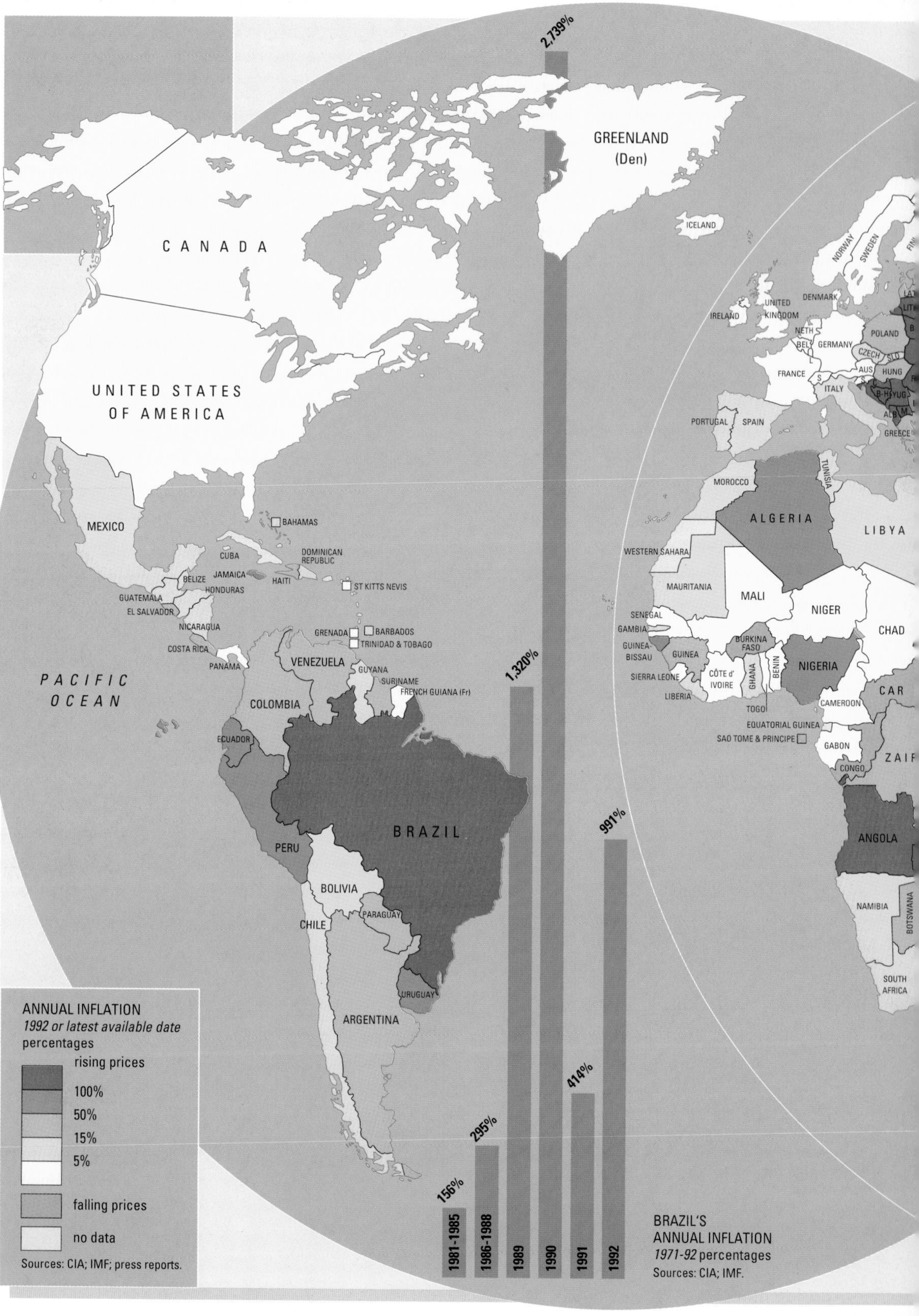

2,739%

GREENLAND
(Den)

ICELAND

CANADA

UNITED STATES
OF AMERICA

NORWAY

SWEDEN

FIN

DENMARK

IRELAND

UNITED
KINGDOM

NETH.

BEL GERMANY POLAND

FRANCE

CZECH
SLO

AUS HUNG

S

ITALY

B-H YUG.

R

ALB M.

PORTUGAL SPAIN

GREECE

MEXICO

BAHAMAS

CUBA

DOMINICAN
REPUBLIC

BELIZE

JAMAICA

HAITI

ST KITTS NEVIS

GUATEMALA

HONDURAS

EL SALVADOR

NICARAGUA

COSTA RICA

PANAMA

GRENADA

BARBADOS

TRINIDAD & TOBAGO

VENEZUELA

GUYANA

SURINAME

FRENCH GUIANA (Fr)

*PACIFIC
OCEAN*

COLOMBIA

ECUADOR

BRAZIL

PERU

BOLIVIA

PARAGUAY

CHILE

URUGUAY

ARGENTINA

MOROCCO

TUNISIA

ALGERIA

LIBYA

WESTERN SAHARA

MAURITANIA

MALI

NIGER

CHAD

SENEGAL

GAMBIA

GUINEA-
BISSAU

GUINEA

BURKINA
FASO

BENIN

NIGERIA

SIERRA LEONE

CÔTE d'
IVOIRE

GHANA

TOGO

CAR

LIBERIA

CAMEROON

EQUATORIAL GUINEA

SAO TOME & PRINCIPE

GABON

CONGO

ZAIR

ANGOLA

NAMIBIA

BOTSWANA

SOUTH
AFRICA

1,320%

991%

414%

156%

295%

1981-1985

1986-1988

1989

1990

1991

1992

ANNUAL INFLATION
1992 or latest available date
percentages

rising prices

100%

50%

15%

5%

falling prices

no data

BRAZIL'S
ANNUAL INFLATION
1971-92 percentages
Sources: CIA; IMF.

Sources: CIA; IMF; press reports.

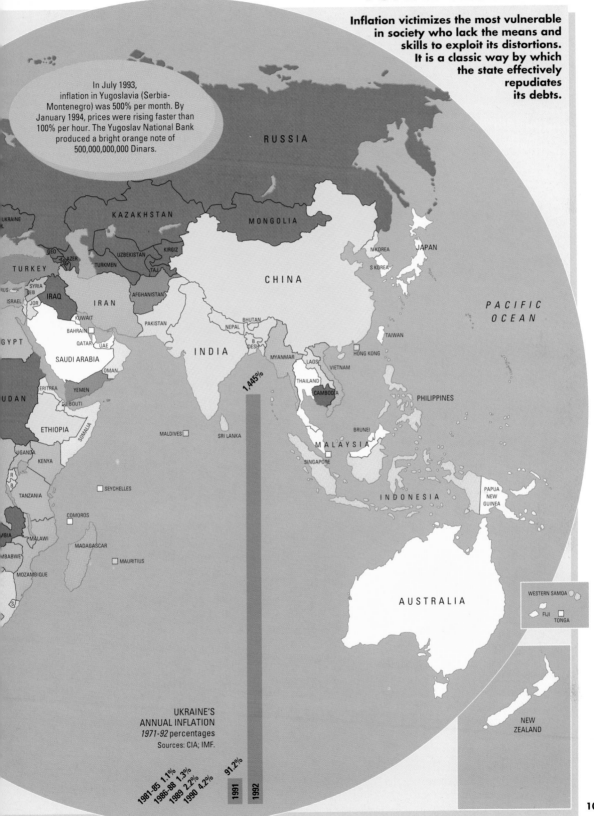

Inflation victimizes the most vulnerable in society who lack the means and skills to exploit its distortions. It is a classic way by which the state effectively repudiates its debts.

In July 1993, inflation in Yugoslavia (Serbia-Montenegro) was 500% per month. By January 1994, prices were rising faster than 100% per hour. The Yugoslav National Bank produced a bright orange note of 500,000,000,000 Dinars.

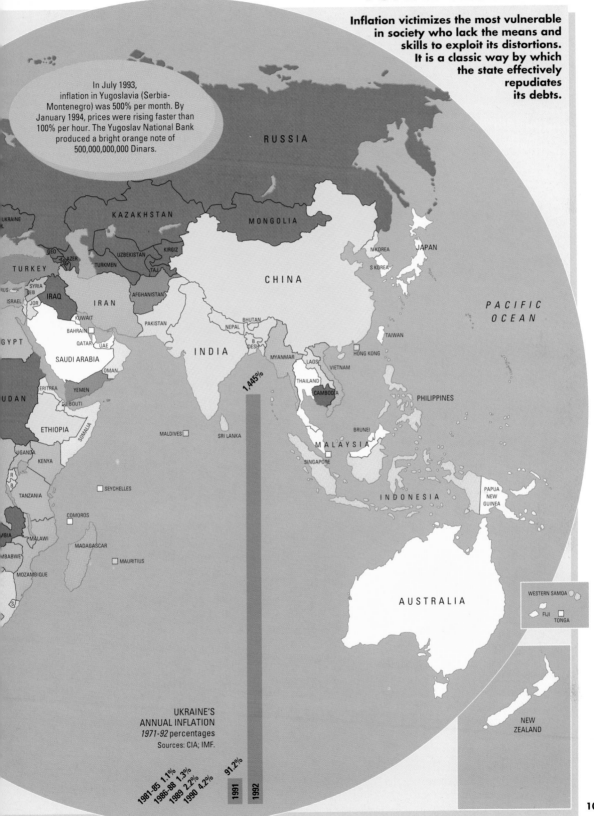

RUSSIA

KAZAKHSTAN

MONGOLIA

UKRAINE

GEO
AZER
UZBEKISTAN
KIRGIZ
TURKMEN
TAJ.

TURKEY

RUS
SYRIA
LEB
ISRAEL
JOR
IRAQ
IRAN
AFGHANISTAN

EGYPT

KUWAIT
BAHRAIN
QATAR
UAE
SAUDI ARABIA
OMAN

PAKISTAN

CHINA

N KOREA
JAPAN
S KOREA

PACIFIC
OCEAN

BHUTAN
NEPAL
B
DESH
INDIA

TAIWAN

HONG KONG

ERITREA
YEMEN
DJIBOUTI
SOMALIA

SUDAN

ETHIOPIA

MYANMAR
LAOS
VIETNAM
THAILAND
CAMBODIA

PHILIPPINES

1,445%

MALDIVES
SRI LANKA

UGANDA
KENYA

BRUNEI

MALAYSIA

SINGAPORE

SEYCHELLES

TANZANIA

INDONESIA

PAPUA
NEW
GUINEA

COMOROS

MBIA
MALAWI

MADAGASCAR

MAURITIUS

MBABWE
MOZAMBIQUE

AUSTRALIA

WESTERN SAMOA

FIJI
TONGA

NEW
ZEALAND

UKRAINE'S
ANNUAL INFLATION
1971-92 percentages
Sources: CIA; IMF.

1981-85 1.1%
1986-88 1.3%
1989 2.2%
1990 4.2%
91.2%
1991
1992

British Sky Broadcasting
Austria
Belgium
Denmark
Finland
France
Israel
Netherlands
Norway
Sweden
Switzerland
UK
and elsewhere

Rupert Murdoch's News Corporation is one of a few
media companies with a considerable international
presence. Others include Time Warner (U.S.A.) and
Bertelsmann (Germany).

Some major media companies are more renowned for
their home dominance but have subsidiary interests
elsewhere. These include Fininvest (Italy), whose
proprietor, Silvio Berlusconi, was for a short period that
country's prime minister; Globo (Brazil); and, Grupo
Televisa (Mexico) the world's leading Spanish-language
media conglomerate.

Sources: Belfield, Hird and Kelly; *The Economist;*
News Corporation Limited; Shawcross;
UN Environment Programme.

Star Television
Afghanistan
China
Hong Kong
India
Indonesia
Jordan
South Korea
Kuwait
Pakistan
Philippines
Saudi Arabia
Taiwan
Thailand
UEA

television
Australia
Luxembourg
U.S.A.

newspapers and magazines
Australia
Fiji
Germany
Hong Kong
New Zealand
Papua New Guinea
Spain
UK
U.S.A.

book publishing
Australia
UK
U.S.A.

film
U.S.A.

CANADA

UNITED STATES
OF AMERICA

ATLANTIC
OCEAN

MEXICO

BAHAMAS

CUBA
DOMINICAN
REPUBLIC
JAMAICA
BELIZE   HAITI   ANTIGUA
GUATEMALA   HONDURAS
EL SALVADOR
NICARAGUA
BARBADOS
COSTA RICA   TRINIDAD & TOBAGO
PANAMA   VENEZUELA   GUYANA
                         SURINAME
                         FRENCH GUIANA (Fr)
COLOMBIA

PACIFIC
OCEAN

ECUADOR

PERU   BRAZIL

BOLIVIA

CHILE   PARAGUAY

URUGUAY

ARGENTINA

ICELAND

NORWAY   SWEDEN   FINLAND

                        ESTONIA
                        LATVIA
IRELAND   LITHUANIA
                DENMARK
UNITED   BELARU
KINGDOM NETH   POLAND
         BEL
              GERMANY CZECH   UKRAIN
                       REPUBLIC
                              SLOVAK
FRANCE   AUSTRIA HUNGARY
         SWITZ   SLO   ROMANIA
                    CROATIA
                  B - H   YUG   BULGARI
PORTUGAL   ALB   M
         SPAIN   ITALY
                                   GREECE

MOROCCO   TUNISIA

WESTERN SAHARA   ALGERIA   LIBYA

         MAURITANIA   MALI   NIGER
CAPE VERDE
         SENEGAL   CHAD
GAMBIA
GUINEA-BISSAU   BURKINA
         GUINEA   FASO   NIGERIA
SIERRA LEONE CÔTE d'   BENIN
              IVOIRE GHANA   CAR
LIBERIA   TOGO CAMEROON
         EQUATORIAL GUINEA
                        GABON   ZAI
                        CONGO
                                ANGOLA

                        NAMIBIA

                        SOUTH
                        AFRICA

STATE ATTITUDES TO RELIGION
early 1990s

states with an official belief

[ ] others repressed

[ ] others tolerated

states with no official belief

[ ] favouritism in practice

[ ] even-handedness in practice

[ ] unknown or unclear

[$] a religion receives substantial
    state funding

Sources: Barrett; O'Brien and Palmer;
personal communications.

110

FUNDAMENTALISM
1994

UNITED STATES
OF AMERICA

MEXICO

CUBA
                    DOMINICAN
                    REPUBLIC
BELIZE HONDURAS   HAITI
GUATEMALA
EL SALVADOR NICARAGUA
COSTA RICA
         PANAMA   VENEZUELA GUYANA
                            SURINAME
                  COLOMBIA FRENCH GUIANA
         ECUADOR

         PERU   BRAZIL

         BOLIVIA

         CHILE   PAR

                    URU

[ ] fundamentalist states

[ ] states challenged by
    fundamentalists

[ ] states where government policy is
    influenced by fundamentalists

[ ] states where government policy takes
    into account fundamentalist opinion

[ ] states where policy is not affected

[ ] unclear or unknown

Sources: Marty and Appleby; press reports;
personal communications.

Beliefs sustain the state. The state tries to sustain belief.

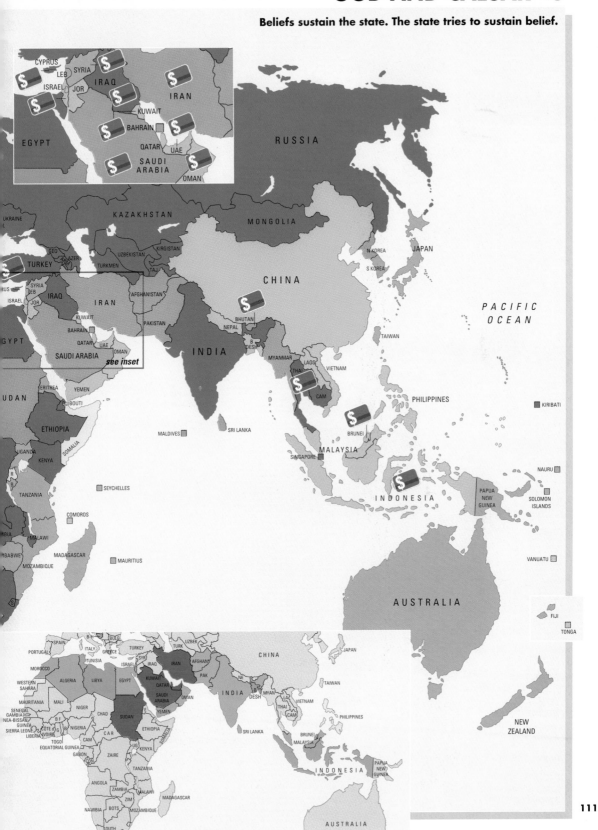

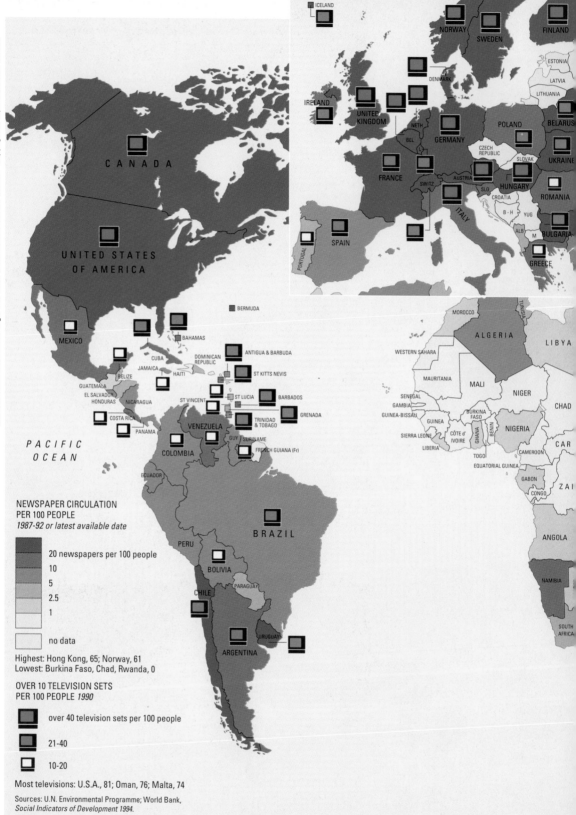

ICELAND

NORWAY

SWEDEN

FINLAND

ESTONIA

LATVIA

LITHUANIA

DENMARK

IRELAND

UNITED KINGDOM

NETH

BEL

GERMANY

POLAND

BELARUS

CZECH REPUBLIC

SLOVAK

UKRAINE

FRANCE

SWITZ

AUSTRIA

HUNGARY

ROMANIA

SLO

CROATIA

B-H   YUG

BULGARIA

ITALY

ALB   M

PORTUGAL

SPAIN

GREECE

C A N A D A

UNITED STATES
OF AMERICA

BERMUDA

MOROCCO

TUNISIA

ALGERIA

LIBYA

MEXICO

BAHAMAS

ANTIGUA & BARBUDA

WESTERN SAHARA

CUBA

DOMINICAN REPUBLIC

ST KITTS NEVIS

MAURITANIA

MALI

NIGER

CHAD

JAMAICA

BELIZE

HAITI

SENEGAL

GAMBIA

BURKINA FASO

BENIN

NIGERIA

GUATEMALA

EL SALVADOR

HONDURAS

NICARAGUA

ST VINCENT

ST LUCIA

BARBADOS

GUINEA-BISSAU

GUINEA

CÔTE d'
IVOIRE

GHANA

COSTA RICA

GRENADA

SIERRA LEONE

CAMEROON

CAR

PANAMA

TRINIDAD
& TOBAGO

LIBERIA

TOGO

VENEZUELA

GUY   SURINAME

EQUATORIAL GUINEA

PACIFIC
OCEAN

COLOMBIA

FRENCH GUIANA (Fr)

GABON

CONGO

ZAI

ECUADOR

NEWSPAPER CIRCULATION
PER 100 PEOPLE
1987-92 or latest available date

PERU

B R A Z I L

ANGOLA

BOLIVIA

20 newspapers per 100 people

10

5

2.5

1

PARAGUAY

NAMIBIA

no data

CHILE

URUGUAY

SOUTH
AFRICA

Highest: Hong Kong, 65; Norway, 61
Lowest: Burkina Faso, Chad, Rwanda, 0

ARGENTINA

OVER 10 TELEVISION SETS
PER 100 PEOPLE *1990*

over 40 television sets per 100 people

21-40

10-20

Most televisions: U.S.A., 81; Oman, 76; Malta, 74

Sources: U.N. Environmental Programme; World Bank,
Social Indicators of Development 1994.

Newspapers are in many states the major instrument for directly influencing political opinion, although television is perhaps the most potent form of mass communication.

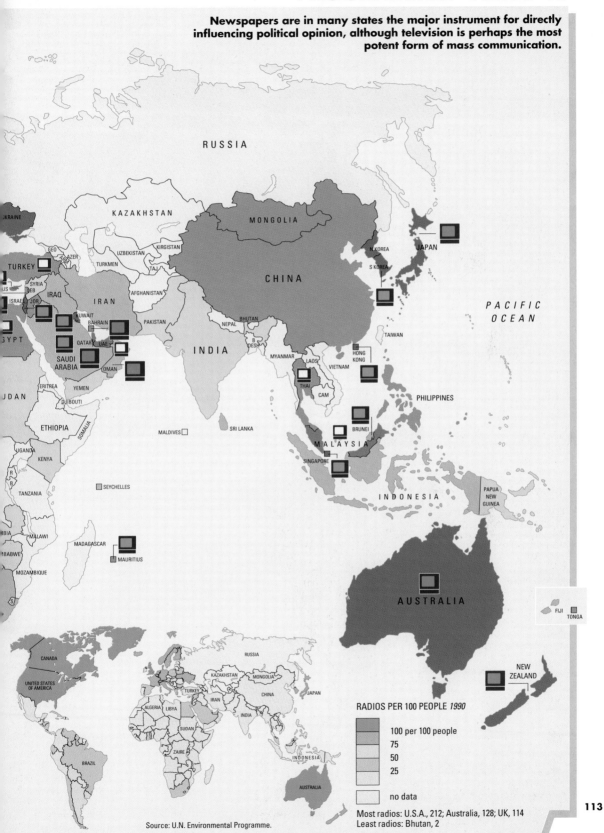

RUSSIA

KAZAKHSTAN

MONGOLIA

JKRAINE

TURKEY

GEO
AZER
UZBEKISTAN
TURKMEN
KIRGISTAN
TAJ

SYRIA
DEB
ISRAEL JOR
IRAQ
IRAN
AFGHANISTAN
CHINA

N.KOREA
JAPAN
S KOREA

EGYPT
KUWAIT
BAHRAIN
QATARY
UAE
OMAN
SAUDI
ARABIA
PAKISTAN
NEPAL
BHUTAN
B
DESH
INDIA

TAIWAN

PACIFIC
OCEAN

JDAN
ERITREA
YEMEN
DJIBOUTI
MYANMAR
LAOS
THAI
CAM
VIETNAM
HONG
KONG

ETHIOPIA
SOMALIA
MALDIVES
SRI LANKA

PHILIPPINES

UGANDA
KENYA
R
B
TANZANIA
SEYCHELLES

MALAYSIA
BRUNEI
SINGAPORE

INDONESIA

PAPUA
NEW
GUINEA

BIA
MALAWI
MADAGASCAR
1BABWE
MOZAMBIQUE
MAURITIUS

AUSTRALIA

FIJI
TONGA

CANADA

RUSSIA

KAZAKHSTAN
MONGOLIA

NEW
ZEALAND

UNITED STATES
OF AMERICA

TURKEY
IRAN
CHINA
JAPAN

ALGERIA
LIBYA
INDIA

SUDAN

ZAIRE

BRAZIL

INDONESIA

AUSTRALIA

RADIOS PER 100 PEOPLE *1990*

100 per 100 people
75
50
25

no data

Most radios: U.S.A., 212; Australia, 128; UK, 114
Least radios: Bhutan, 2

Source: U.N. Environmental Programme.

GREENLAND
(Den)

ICELAND

C A N A D A

NORWAY
SWEDEN

DENMARK

IRELAND
UNITED
KINGDOM

NETH.
BEL.
GERMANY
POLAND
CZECH
SLO

UNITED STATES
OF AMERICA

FRANCE
AUS.
HUNG
S
R-HVUG
ITALY
ALB.

PORTUGAL
SPAIN

MALTA

BAHAMAS

ATLANTIC
OCEAN

MOROCCO

TUNISIA

ALGERIA

LIBYA

MEXICO

WESTERN SAHARA

CUBA

DOMINICAN
REPUBLIC

BELIZE
JAMAICA
HONDURAS
HAITI
ST. KITTS NEVIS
ANGUILLA

ANTIGUA AND
BARBUDA

MAURITANIA

MALI

NIGER

CHAD

GUATEMALA
EL SALVADOR

DOMINICA

SENEGAL
GAMBIA
GUINEA-BISSAU

BURKINA
FASO

NICARAGUA

ST. VINCENT &
GRENADINES
GRENADA

ST. LUCIA

GUINEA

BENIN

NIGERIA

COSTA RICA

TRINIDAD & TOBAGO

SIERRA LEONE

CÔTE d'
IVOIRE
GHANA

C A R

PANAMA

VENEZUELA

GUYANA
SURINAME

LIBERIA

TOGO

FRENCH GUIANA (Fr)

EQUATORIAL GUINEA

CAMEROON

COLOMBIA

SAO TOME & PRINCIPE

GABON

Z A

ECUADOR

CONGO

B R A Z I L

ANGOLA

PERU

CENSORSHIP:
STATES' INFRINGEMENTS OF
FREEDOMS OF BELIEF, EXPRESSION,
COMMUNICATION AND MOVEMENT
early 1990s

BOLIVIA

PARAGUAY

NAMIBIA

CHILE

draconian
*imprisonment, or worse, for holding
views hostile to the state or its
representatives, or withholding public
support for them*

SOUTH
AFRICA

URUGUAY

severe
*punishment is common for expressing
hostility to the state or its representatives*

ARGENTINA

mild
*laws governing the expression of opinion
are usually in abeyance, except in times
of war*

unclear or unknown

situation worsening *end 1994*

situation improving *end 1994*

Sources: *Index on Censorship;* U.S. State Department.

All states abridge their citizens' freedoms: to hold opinions, to express them, and to travel. Some set the limits wide, some narrow.

RUSSIA

KAZAKHSTAN

MONGOLIA

UKRAINE

GEO
AZER
UZBEKISTAN
KIRGISTAN
TURKEY
TURKMEN
TAJ

N.KOREA
JAPAN

S.KOREA

CHINA

PACIFIC
OCEAN

CPRUS
SYRIA
LEB
ISRAEL
JOR
IRAQ
IRAN
AFGHANISTAN
KUWAIT
PAKISTAN

BHUTAN
NEPAL
TAIWAN

BAHRAIN
QATAR UAE
B
DESH
HONG KONG

EGYPT
SAUDI ARABIA
OMAN
INDIA
MYANMAR
LAOS
VIETNAM

SUDAN
ERITREA
YEMEN
DJIBOUTI
THAILAND
CAM

PHILIPPINES

KIRIBATI

ETHIOPIA
MALDIVES
SRI LANKA
MARSHALL
ISLANDS

UGANDA
KENYA
BRUNEI
MALAYSIA
NAURU

BURUNDI
TANZANIA
SEYCHELLES
SINGAPORE

INDONESIA
PAPUA
NEW
GUINEA
SOLOMON
ISLANDS

COMOROS

AMBIA
MALAWI
MADAGASCAR
MAURITIUS

PRISONERS OF CONSCIENCE
January 1994

Amnesty International prisoners of conscience and/or International PEN cases

other states

ZIMBABWE

MOZAMBIQUE

AUSTRALIA

WESTERN SAMOA

FIJI
TONGA

LESOTHO

GREENLAND
(Den)

CANADA

RUSSIA

UNITED STATES
OF AMERICA
KAZAKHSTAN
MONGOLIA
TURKEY
CHINA
JAPAN
IRAN
ALGERIA LIBYA
INDIA
SUDAN
NEW
ZEALAND

BRAZIL
ZAIRE

INDONESIA

AUSTRALIA

Sources: Amnesty International; International PEN.

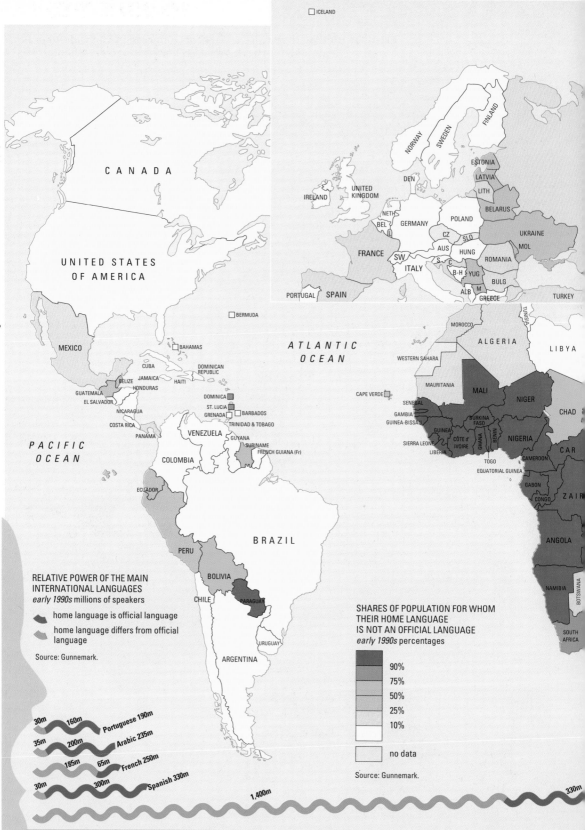

CANADA

UNITED STATES
OF AMERICA

MEXICO

GUATEMALA
EL SALVADOR
BELIZE
HONDURAS
NICARAGUA
COSTA RICA
PANAMA

BERMUDA

BAHAMAS

CUBA
JAMAICA
HAITI
DOMINICAN
REPUBLIC
DOMINICA
ST. LUCIA
GRENADA
BARBADOS
TRINIDAD & TOBAGO

VENEZUELA
COLOMBIA
ECUADOR
GUYANA
SURINAME
FRENCH GUIANA (Fr)

PERU
BRAZIL
BOLIVIA
CHILE
PARAGUAY
URUGUAY
ARGENTINA

PACIFIC
OCEAN

ATLANTIC
OCEAN

ICELAND

NORWAY
SWEDEN
FINLAND
DEN
ESTONIA
LATVIA
LITH
BELARUS
IRELAND
UNITED
KINGDOM
NETH
BEL
GERMANY
POLAND
UKRAINE
CZ
SLO
MOL
FRANCE
SW
AUS
HUNG
ROMANIA
ITALY
S
C
B-H
YUG
BULG
ALB
M
PORTUGAL
SPAIN
GREECE
TURKEY

MOROCCO
TUNISIA
ALGERIA
LIBYA
WESTERN SAHARA
MAURITANIA
CAPE VERDE
MALI
NIGER
CHAD
SENEGAL
GAMBIA
GUINEA-BISSAU
BURKINA
FASO
GUINEA
SIERRA LEONE
LIBERIA
CÔTE d'
IVOIRE
GHANA
BENIN
TOGO
NIGERIA
CAMEROON
EQUATORIAL GUINEA
GABON
CONGO
CAR
ZAIRE
ANGOLA
NAMIBIA
BOTSWANA
SOUTH
AFRICA

RELATIVE POWER OF THE MAIN
INTERNATIONAL LANGUAGES
early 1990s millions of speakers

home language is official language

home language differs from official
language

Source: Gunnemark.

30m   160m
Portuguese 190m
35m   200m
Arabic 235m
185m   65m
French 250m
30m   300m
Spanish 330m

1,400m

330m

SHARES OF POPULATION FOR WHOM
THEIR HOME LANGUAGE
IS NOT AN OFFICIAL LANGUAGE
early 1990s percentages

90%
75%
50%
25%
10%

no data

Source: Gunnemark.

116

Only 80 of the world's 5,300 or so languages are official languages.

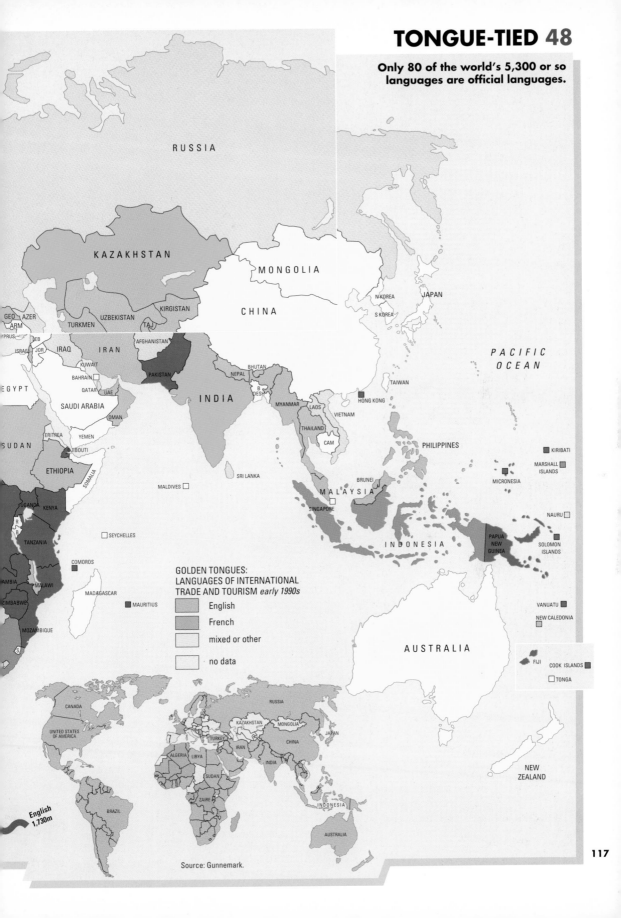

RUSSIA

KAZAKHSTAN

MONGOLIA

CHINA

GEO AZER
ARM
CYPRUS
ISRAEL JOR
JEB
IRAQ
IRAN
AFGHANISTAN
TURKMEN
UZBEKISTAN
TAJ
KIRGISTAN

N KOREA
JAPAN
S KOREA

PACIFIC
OCEAN

EGYPT
KUWAIT
BAHRAIN
QATAR UAE
SAUDI ARABIA
OMAN
YEMEN
ERITREA
SUDAN
DJIBOUTI
ETHIOPIA
SOMALIA

PAKISTAN
NEPAL
BHUTAN
INDIA
B
DESH
MYANMAR
LAOS
THAILAND
VIETNAM
CAM

TAIWAN

HONG KONG

PHILIPPINES

KIRIBATI

MARSHALL
ISLANDS

MICRONESIA

SRI LANKA
MALDIVES

UGANDA KENYA
R
B
TANZANIA
SEYCHELLES

BRUNEI
MALAYSIA
SINGAPORE

INDONESIA

NAURU

PAPUA
NEW
GUINEA
SOLOMON
ISLANDS

COMOROS
MADAGASCAR
MAURITIUS
AMBIA
MALAWI
ZIMBABWE
MOZAMBIQUE
S

**GOLDEN TONGUES:
LANGUAGES OF INTERNATIONAL
TRADE AND TOURISM** *early 1990s*

- English
- French
- mixed or other
- no data

VANUATU
NEW CALEDONIA

AUSTRALIA

FIJI
COOK ISLANDS
TONGA

CANADA

UNITED STATES
OF AMERICA

RUSSIA
KAZAKHSTAN
MONGOLIA
TURKEY
CHINA
JAPAN
ALGERIA LIBYA
IRAN
INDIA
SUDAN
ZAIRE
BRAZIL
INDONESIA
AUSTRALIA

NEW
ZEALAND

English
1,730m

117

Source: Gunnemark.

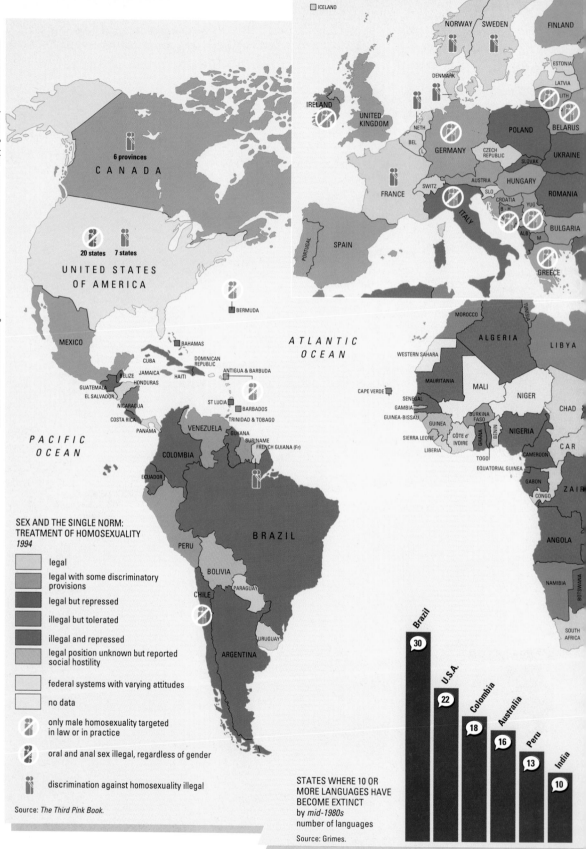

6 provinces

CANADA

20 states   7 states

UNITED STATES
OF AMERICA

MEXICO

BERMUDA

PACIFIC
OCEAN

CUBA
BAHAMAS
JAMAICA   DOMINICAN
HAITI   REPUBLIC
BELIZE
GUATEMALA   HONDURAS
EL SALVADOR
NICARAGUA
COSTA RICA
PANAMA

ANTIGUA & BARBUDA

ST LUCIA
BARBADOS
TRINIDAD & TOBAGO

VENEZUELA
GUYANA
SURINAME
FRENCH GUIANA (Fr)
COLOMBIA
ECUADOR

PERU

BOLIVIA

CHILE   PARAGUAY

URUGUAY
ARGENTINA

BRAZIL

ATLANTIC
OCEAN

ICELAND

NORWAY   SWEDEN   FINLAND

DENMARK   ESTONIA
LATVIA
LITH
IRELAND
UNITED
KINGDOM   BELARUS
NETH   POLAND
BEL   GERMANY   UKRAINE
CZECH
REPUBLIC
SLOVAK
FRANCE   SWITZ   AUSTRIA   HUNGARY
SLO   ROMANIA
CROATIA
B-H   YUG   BULGARIA
ITALY   ALB   M
PORTUGAL
SPAIN   GREECE

MOROCCO   TUNISIA
ALGERIA   LIBYA
WESTERN SAHARA
MAURITANIA   MALI   NIGER   CHAD
CAPE VERDE
SENEGAL
GAMBIA
GUINEA-BISSAU   BURKINA
GUINEA   FASO   NIGERIA
SIERRA LEONE   CÔTE d'   BENIN
IVOIRE   GHANA   CAR
LIBERIA   TOGO   CAMEROON
EQUATORIAL GUINEA
GABON   ZAIR
CONGO

ANGOLA

NAMIBIA   BOTSWANA

SOUTH
AFRICA

SEX AND THE SINGLE NORM:
TREATMENT OF HOMOSEXUALITY
1994

- legal
- legal with some discriminatory provisions
- legal but repressed
- illegal but tolerated
- illegal and repressed
- legal position unknown but reported social hostility

- federal systems with varying attitudes
- no data

- only male homosexuality targeted in law or in practice
- oral and anal sex illegal, regardless of gender
- discrimination against homosexuality illegal

Source: *The Third Pink Book.*

STATES WHERE 10 OR
MORE LANGUAGES HAVE
BECOME EXTINCT
by *mid-1980s*
number of languages

Brazil **30**
U.S.A. **22**
Colombia **18**
Australia **16**
Peru **13**
India **10**

Source: Grimes.

States want their citizens to conform. Sexual expression and language are just two areas in which societies seek to suppress differences.

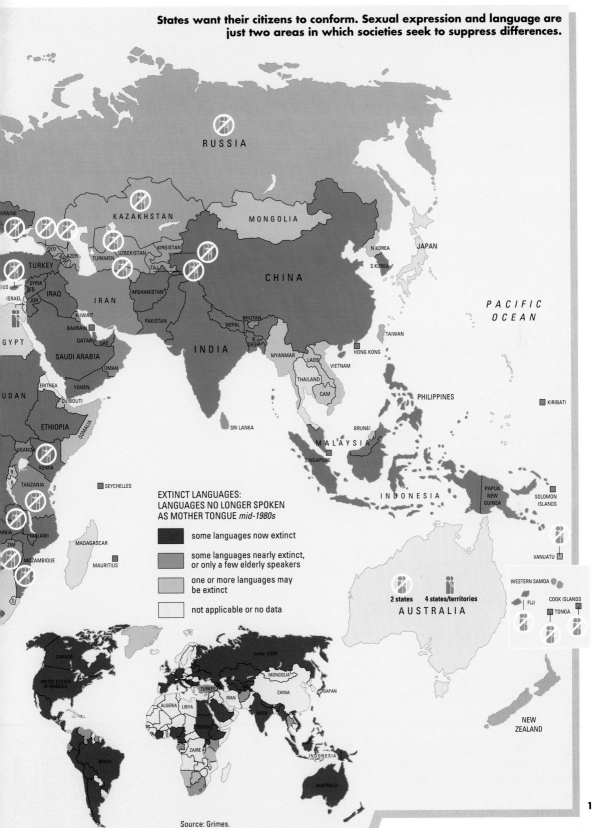

RUSSIA

KAZAKHSTAN

MONGOLIA

KRAINE

GEO
AZER
TURKMEN UZBEKISTAN
KIRGISTAN
TURKEY
TAJ.
N KOREA
JAPAN
S KOREA

CHINA

CYPRUS
SYRIA
LEB
ISRAEL
JOR
IRAQ

IRAN
AFGHANISTAN

GYPT

KUWAIT

BAHRAIN

QATAR
UAE

SAUDI ARABIA

OMAN

PAKISTAN

NEPAL
BHUTAN
B
DESH

INDIA

MYANMAR
LAOS
VIETNAM

TAIWAN

PACIFIC
OCEAN

ERITREA

YEMEN

THAILAND

CAM

HONG KONG

UDAN

DJIBOUTI

PHILIPPINES

KIRIBATI

ETHIOPIA

SOMALIA

SRI LANKA

BRUNEI

UGANDA

KENYA

MALAYSIA

SINGAPORE

R

TANZANIA

SEYCHELLES

INDONESIA

PAPUA
NEW
GUINEA

SOLOMON
ISLANDS

MBIA
MALAWI

MADAGASCAR

ZIM
MOZAMBIQUE

MAURITIUS

VANUATU

S

**EXTINCT LANGUAGES:
LANGUAGES NO LONGER SPOKEN
AS MOTHER TONGUE** *mid-1980s*

some languages now extinct

some languages nearly extinct,
or only a few elderly speakers

one or more languages may
be extinct

not applicable or no data

2 states 4 states/territories

AUSTRALIA

WESTERN SAMOA

FIJI

COOK ISLANDS

TONGA

NEW
ZEALAND

CANADA

former USSR

MONGOLIA

UNITED STATES
OF AMERICA

TURKEY
IRAN

CHINA

JAPAN

ALGERIA LIBYA

INDIA

SUDAN

ZAIRE

BRAZIL

INDONESIA

AUSTRALIA

Source: Grimes.

Even key social indices, such as mortality rates of children under five years old, numbers of malnourished children, and children in primary education, are for many states likely to be inadequate or dated. The bar chart shows — for 100 states in Africa, Asia, Central and South America, and the Caribbean — the age of data internationally available for these indices in 1993.

Poor states are more likely than rich ones to lack the necessary resources to collect and organize information regularly. Government policy may not rate the provision of data as important enough for the investment of scarce resources. Indeed, governments may sometimes refuse to make such data public, regarding it as likely to prove more embarrassing than useful.

Statistics may be up to date but still unreliable. By their very selectiveness, in rich states as well as poor ones, statistics may mislead rather than reveal.

Source: UNICEF, *The Progress of Nations*, 1994.

45 states

28 states

18 states

4 states

3-4 years old   5-9 years old   10-14 years old   15 years old

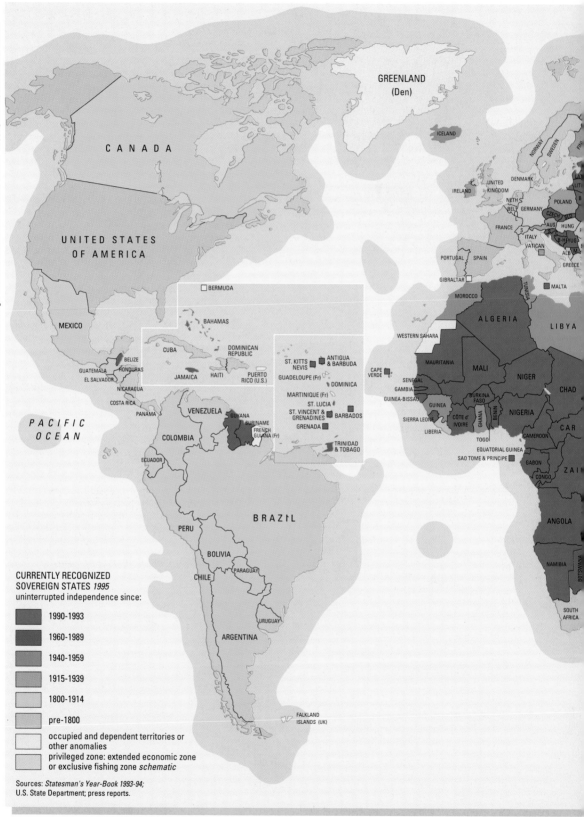

GREENLAND
(Den)

ICELAND

CANADA

NORWAY
SWEDEN

UNITED STATES
OF AMERICA

DENMARK
IRELAND
UNITED
KINGDOM
NETH
BEL
GERMANY
POLAND
CZECH
SL
FRANCE
AUS
HUNG
B-H
YUG
ITALY
VATICAN
ALB
PORTUGAL
SPAIN
GREECE

BERMUDA

GIBRALTAR
MALTA

MEXICO

BAHAMAS

MOROCCO
TUNISIA

ALGERIA
LIBYA

WESTERN SAHARA

CUBA

DOMINICAN
REPUBLIC

BELIZE
GUATEMALA HONDURAS
EL SALVADOR
NICARAGUA

JAMAICA
HAITI

ST. KITTS
NEVIS

PUERTO
RICO (U.S.)

ANTIGUA
& BARBUDA

GUADELOUPE (Fr)
DOMINICA

MAURITANIA
MALI
NIGER
CHAD

CAPE
VERDE
SENEGAL
GAMBIA
GUINEA-BISSAU

MARTINIQUE (Fr)
ST. LUCIA

BURKINA
FASO
GUINEA

COSTA RICA
PANAMA

VENEZUELA

GUYANA
SURINAME
FRENCH
GUIANA (Fr)

ST. VINCENT &
GRENADINES BARBADOS
GRENADA

SIERRA LEONE
LIBERIA

CÔTE d'
IVOIRE
GHANA
BENIN
TOGO

NIGERIA
CAMEROON
CAR

PACIFIC
OCEAN

COLOMBIA

TRINIDAD
& TOBAGO

EQUATORIAL GUINEA
SAO TOME & PRINCIPE

GABON
ZAI
CONGO

ECUADOR

BRAZIL

ANGOLA

PERU

BOLIVIA

PARAGUAY

NAMIBIA
BOTSWANA

CHILE

SOUTH
AFRICA

URUGUAY

ARGENTINA

FALKLAND
ISLANDS (UK)

**CURRENTLY RECOGNIZED
SOVEREIGN STATES** *1995*
uninterrupted independence since:

- 1990-1993
- 1960-1989
- 1940-1959
- 1915-1939
- 1800-1914
- pre-1800
- occupied and dependent territories or other anomalies
- privileged zone: extended economic zone or exclusive fishing zone *schematic*

Sources: *Statesman's Year-Book 1993-94;*
U.S. State Department; press reports.

States exist only in each other's recognition, yet they divide the earth and its waters between them. Big or small they pretend to permanence, yet are a blink in history.

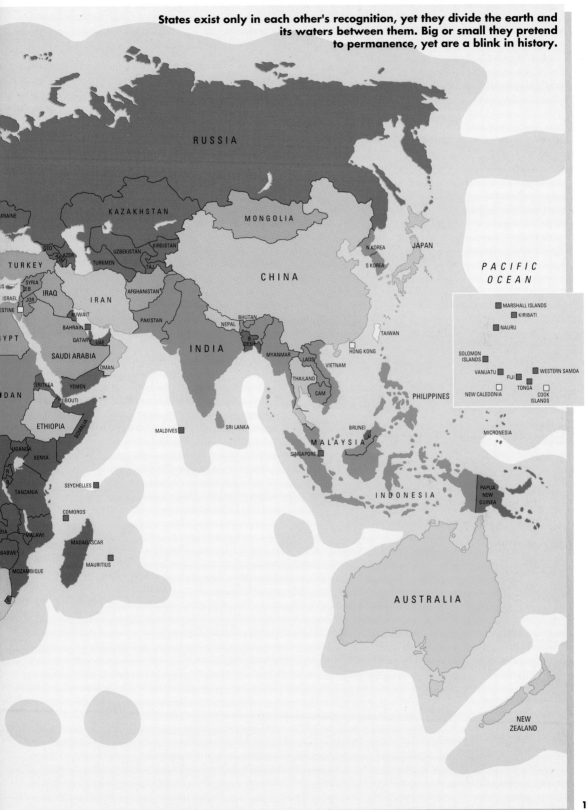

RUSSIA

KAZAKHSTAN

MONGOLIA

KRAINE

GEO

AZER

UZBEKISTAN

KIRGISTAN

N KOREA

JAPAN

TURKEY

TURKMEN

TAJ

S KOREA

PACIFIC OCEAN

JS

SYRIA

DEB

AFGHANISTAN

CHINA

ISRAEL

JOR

IRAQ

IRAN

ESTINE

KUWAIT

PAKISTAN

BHUTAN

TAIWAN

YPT

BAHRAIN

NEPAL

QATAR

UAE

B DESH

HONG KONG

MARSHALL ISLANDS

SAUDI ARABIA

OMAN

INDIA

MYANMAR

LAOS

VIETNAM

KIRIBATI

NAURU

DAN

ERITREA

YEMEN

THAILAND

CAM

SOLOMON ISLANDS

ETHIOPIA

DJIBOUTI

SOMALIA

PHILIPPINES

VANUATU

FIJI

WESTERN SAMOA

UGANDA

KENYA

MALDIVES

SRI LANKA

BRUNEI

MICRONESIA

NEW CALEDONIA

TONGA

COOK ISLANDS

R

TANZANIA

SEYCHELLES

MALAYSIA

SINGAPORE

COMOROS

IA

MALAWI

MADAGASCAR

INDONESIA

PAPUA NEW GUINEA

BABWE

MOZAMBIQUE

MAURITIUS

AUSTRALIA

NEW ZEALAND

	1 International frontiers 1994 kms		2 Population mid-1993 millions	3 Young and old people 1994 percentage of population		4 Populatio density 1993 people per kr
	land	**coastline**		**under 15 years**	**over 65 years**	
Afghanistan	5,529	0	17.4	46%	4%	32.2
Albania	720	362	3.3	33%	5%	114.8
Algeria	6,343	998	27.3	44%	4%	10.8
Angola	5,198	1,600	9.5	45%	3%	7.6
Argentina	9,665	4,989	33.5	30%	9%	11.8
Armenia	1,254	0	3.6	30%	5%	120.3
Australia	0	25,760	17.8	22%	11%	2.2
Austria	2,496	0	7.9	18%	15%	93.3
Azerbaijan	2,013	0	7.2	33%	5%	83.6
Bahamas	0	3,542	0.3	30%	5%	18.7
Bahrain	0	161	0.5	35%	2%	758.8
Bangladesh	4,246	580	113.9	44%	3%	778.8
Barbados	0	97	0.3	25%	11%	600.2
Belarus	3,098	0	10.3	23%	10%	49.5
Belgium	1,385	64	10.1	18%	15%	–
Belize	516	386	0.2	44%	5%	8.5
Benin	1,989	121	5.1	46%	3%	43.4
Bhutan	1,878	0	0.8	39%	4%	31.2
Bolivia	6,743	0	8.0	41%	4%	6.7
Bosnia-Herzegovina	1,459	20	4.0	28%	6%	–
Botswana	4,013	0	1.4	48%	3%	2.3
Brazil	14,691	7,491	152.0	35%	5%	17.8
Brunei	381	161	0.3	36%	3%	45.9
Bulgaria	1,808	354	9.0	20%	13%	77.3
Burkina Faso	3,192	0	10.0	48%	4%	33.8
Burundi	974	0	5.8	46%	4%	203.1
Cambodia	2,572	443	9.0	35%	3%	48.8
Cameroon	4,591	402	12.8	45%	3%	25
Canada	8,893	243,791	28.1	21%	12%	2.7
Central African Rep.	5,203	0	3.1	42%	3%	5
Chad	5,968	0	5.4	41%	4%	4.5
Chile	6,171	6,435	13.5	31%	6%	17.7
China	22,143	14,500	1,178.5	28%	6%	120.1
Colombia	7,408	3,208	34.9	34%	4%	28.8
Comoros	0	340	0.5	48%	3%	220.6
Congo	5,504	169	2.4	44%	3%	6.9
Costa Rica	639	1,290	3.3	37%	5%	60.9
Côte d'Ivoire	3,110	515	13.4	48%	3%	38.5
Croatia	2,028	5,790	4.4	21%	12%	84.7
Cuba	29	3,735	11.0	23%	9%	96.8
Cyprus	0	648	0.7	26%	10%	76.8
Czech Republic	1,880	0	10.3	21%	13%	–
Denmark	68	3,379	5.2	17%	16%	119.6

Sources: Col 1: *CIA, World Factbook 1994;* **Cols 2 and 3:** Population Concern, *1993 World Population Data Sheet;* **Cols 4 and 5:** World Bank, *Social Indicators of Development 1994;* **Col 6:** *World Bank Atlas 1995;* **Col 7:** Krishna Ahooja-Patel, 1993. The index is based on seven indicators: income,

5 mber of people per hospital bed 1994	6 Purchasing power per person 1993 U.S. dollars	7 Women's advancement 1993 index (1 = best)	8 Number of cars 1993 thousands	9 Number of letters and cards posted per person 1992	
4,003	–	107	55.0	–	Afghanistan
246	340	–	48.9	–	Albania
400	1,650	70	1,535.0	12.72	Algeria
771	–	–	147.5	0.08	Angola
216	7,290	28	5,835.0	0.59	Argentina
120	660	–	14.8	–	Armenia
183	17,510	–	9,954.5	195.23	Australia
94	23,120	–	3,563.0	–	Austria
101	730	–	355.2	–	Azerbaijan
227	11,500	–	–	–	Bahamas
302	7,870	–	140.3	–	Bahrain
3,158	220	97	99.0	1.43	Bangladesh
117	6,240	–	–	34.79	Barbados
8	2,840	–	–	–	Belarus
121	21,210	13	4,482.6	123.31	Belgium
185	2,440	–	–	–	Belize
886	420	77	34.0	0.76	Benin
–	170	102	–	0.95	Bhutan
828	770	87	230.0	0.86	Bolivia
–	–	–	–	–	Bosnia-Herzegovina
421	2,590	42	71.6	–	Botswana
301	3,020	36	12,950.0	18.12	Brazil
319	–	–	128.4	26.11	Brunei
100	1,160	–	1,489.0	24.03	Bulgaria
3,392	300	88	23.5	1.20	Burkina Faso
890	180	68	18.0	0.44	Burundi
478	–	108	30.0	–	Cambodia
393	770	48	165.0	–	Cameroon
64	20,670	3	17,014.5	–	Canada
1,171	390	91	20.0	–	Central African Rep.
1,373	200	105	15.0	0.30	Chad
320	3,070	24	1,167.7	17.81	Chile
388	490	23	6,988.8	4.88	China
703	1,400	38	1,470.0	–	Colombia
476	520	–	–	–	Comoros
228	920	64	46.0	–	Congo
302	2,160	26	178.9	3.97	Costa Rica
1,269	630	66	255.0	2.78	Côte d'Ivoire
–	–	–	–	–	Croatia
207	–	–	47.0	–	Cuba
180	10,380	–	298.8	–	Cyprus
–	2,730	–	3,460.7	110.04	Czech Republic
177	26,510	2	1,917.0	214.87	Denmark

maternal mortality, infant mortality, life expectancy, post-secondary school enrolment, share of paid employment and participation in national parliament;
Col 8: Motor Industry of Great Britain, *World Resources 1994-95;* **Col 9:** Union Postale Universelle, *Statistique des services postaux 1992.*

	1 International Frontiers 1994 kms		2 Population mid-1993 millions	3 Young and old people 1994 percentage of population		4 Populatio density 1993
	land	coastline		under 15 years	over 65 years	people per k
Djibouti	508	314	0.5	45%	3%	22.5
Dominican Republic	275	1,288	7.6	38%	3%	147.7
Ecuador	2,010	2,237	10.3	39%	4%	38
Egypt	2,689	3,450	58.3	39%	4%	53.5
El Salvador	545	307	5.2	45%	4%	250.9
Equatorial Guinea	539	296	0.4	43%	4%	15.2
Eritrea	1,630	1,151	–	–	–	–
Estonia	557	1,393	1.6	22%	12%	34.7
Ethiopia	5,311	0	56.7	49%	4%	43.3
Fiji	0	1,129	0.8	38%	3%	40.6
Finland	2,628	1,126	5.1	19%	14%	14.8
France	2,892	3,427	57.7	20%	14%	103.5
Gabon	2,551	885	1.1	33%	6%	4.4
Gambia	740	80	0.9	44%	3%	84.8
Georgia	1,461	310	5.5	25%	9%	78.6
Germany	3,621	2,389	81.1	16%	15%	224.1
Ghana	2,093	539	16.4	45%	3%	64.2
Greece	1,210	13,676	10.5	19%	14%	77.3
Guatemala	1,687	400	10.0	45%	3%	86.9
Guinea	3,399	320	6.2	44%	3%	24.1
Guinea Bissau	724	350	1.0	43%	3%	27.7
Guyana	2,462	459	0.8	35%	4%	3.7
Haiti	275	1,771	6.5	40%	4%	237.6
Honduras	1,520	820	5.6	47%	4%	46.9
Hong Kong	30	733	5.8	21%	9%	5,533.7
Hungary	1,989	0	10.3	20%	13%	111.2
Iceland	0	4,988	0.3	25%	11%	2.5
India	14,103	7,000	897.4	36%	4%	263.6
Indonesia	2,602	54,716	187.6	37%	4%	95.2
Iran	5,440	2,440	62.8	47%	3%	35
Iraq	3,631	58	19.2	48%	3%	42.4
Ireland	360	1,448	3.6	27%	11%	50.1
Israel	1,006	273	5.3	31%	9%	238.9
Italy	1,899	4,996	57.8	17%	14%	191.7
Jamaica	0	1,022	2.4	33%	8%	216.2
Japan	0	29,791	124.8	18%	13%	328
Jordan	1,619	26	3.8	44%	3%	41.1
Kazakhstan	12,012	0	17.2	32%	6%	6.2
Kenya	3,446	536	27.7	49%	2%	43
Kirgistan	3,878	0	4.6	37%	5%	22.4
Korea (North)	1,673	2,495	22.6	29%	4%	184.1
Korea (South)	238	2,413	44.6	26%	5%	437
Kuwait	464	499	1.7	42%	1%	81.6

Sources: Col 1: *CIA, World Factbook 1994;* **Cols 2 and 3:** Population Concern, *1993 World Population Data Sheet;* **Cols 4 and 5:** World Bank, *Social Indicators of Development 1994;* **Col 6:** *World Bank Atlas1995;* **Col 7:** Krishna Ahooja-Patel, 1993. The index is based on seven indicators: income,

WORLD TABLE

5 umber of people per hospital bed 1994	6 Purchasing power per person 1993 U.S. dollars	7 Women's advancement 1993 index (1= best)	8 Number of cars 1993 thousands	9 Number of letters and cards posted per person 1992	
265	780	–	15.0	–	Djibouti
529	1,080	58	245.0	–	Dominican Republic
625	1,170	57	240.0	0.78	Ecuador
536	660	78	1,585.0	2.88	Egypt
699	1,320	55	255.1	1.11	El Salvador
–	360	–	–	–	Equatorial Guinea
–	–	–	–	–	Eritrea
82	3,040	–	366.4	–	Estonia
4,141	100	98	52.5	–	Ethiopia
364	2,140	–	65.0	26.05	Fiji
93	18,970	6	2,230.6	81.89	Finland
109	22,360	5	29,060.0	210.25	France
800	4,050	47	40.0	–	Gabon
613	360	–	8.0	–	Gambia
90	560	–	–	–	Georgia
118	23,560	18	42,009.6	99.41	Germany
685	430	84	124.0	5.13	Ghana
196	7,390	–	2,533.3	31.74	Greece
673	1,110	51	295.0	5.99	Guatemala
1,816	510	99	27.5	–	Guinea
538	220	–	5.0	–	Guinea Bissau
300	350	–	31.0	–	Guyana
1,323	–	65	50.0	26.86	Haiti
993	580	49	77.5	–	Honduras
234	17,860	12	451.3	111.86	Hong Kong
99	3,330	–	2,234.5	82.31	Hungary
–	23,620	–	136.1	122.76	Iceland
1,371	290	76	4,667.7	12.82	India
1,503	730	72	3,088.7	1.73	Indonesia
724	2,230	81	2,100.6	–	Iran
603	–	44	775.0	1.61	Iraq
101	12,580	15	1,013.6	123.89	Ireland
219	13,760	16	1,060.0	73.91	Israel
133	19,620	10	32,347.42	59.05	Italy
293	1,390	31	112.5	20.50	Jamaica
64	31,450	14	61,658.1	190.10	Japan
519	1,190	52	248.3	9.41	Jordan
75	1,540	–	–	–	Kazakhstan
623	270	61	310.0	14.05	Kenya
85	830	–	–	–	Kirgistan
–	–	–	–	–	Korea (North)
300	7,670	34	5,231.0	50.36	Korea (South)
241	23,350	20	560.0	–	Kuwait

maternal mortality, infant mortality, life expectancy, post-secondary school enrolment, share of paid employment and participation in national parliament;
Col 8: Motor Industry of Great Britain, *World Resources 1994-95*; **Col 9:** Union Postale Universelle, *Statistique des services postaux 1992*.

| | International frontiers 1994 kms | | Population mid-1993 millions | Young and old people 1994 percentage of population | | Population density 1993 people per km |
| | **1** | | **2** | **3** | | **4** |
	land	coastline		under 15 years	over 65 years	
Laos	5,803	0	4.6	45%	4%	18
Latvia	1,078	531	2.6	21%	12%	41.3
Lebanon	454	225	3.6	40%	5%	356.5
Lesotho	909	0	1.9	41%	4%	59.7
Liberia	1,585	579	2.8	46%	4%	24.2
Libya	4,383	1,770	4.2	50%	2%	2.7
Lithuania	1,273	108	3.8	23%	11%	57.6
Luxembourg	359	0	0.4	17%	13%	–
Macedonia	748	0	2.0	29%	7%	–
Madagascar	0	4,828	13.3	47%	3%	20.5
Malawi	2,881	0	10.0	48%	3%	74.2
Malaysia	2,669	4,675	18.4	37%	4%	55.1
Mali	7,243	0	8.9	46%	4%	7
Malta	0	140	0.4	23%	11%	1,115.6
Marshall Islands	0	370	0.1	51%	3%	–
Mauritania	5,074	754	2.2	44%	4%	2
Mauritius	0	177	1.1	30%	5%	532.8
Mexico	4,538	9,330	90.0	38%	4%	42.5
Moldova	1,389	0	4.4	28%	8%	129.5
Mongolia	8,114	0	2.3	44%	4%	1.4
Morocco	2,002	1,835	28.0	40%	4%	57.4
Mozambique	4,571	2,470	15.3	44%	3%	20.1
Myanmar	5,876	1,930	43.5	37%	4%	63.2
Namibia	3,824	1,572	1.6	46%	3%	1.8
Nepal	2,926	0	20.4	42%	3%	137.8
Netherlands	1,027	451	15.2	18%	13%	403.7
New Zealand	0	15,134	3.4	23%	12%	12.6
Nicaragua	1,231	910	4.1	46%	3%	29
Niger	5,697	0	8.5	49%	3%	6.2
Nigeria	4,047	853	95.1	45%	3%	107.2
Norway	2,515	21,925	4.3	20%	16%	13.2
Oman	1,374	2,092	1.6	47%	3%	7.5
Pakistan	6,774	1,046	122.4	44%	4%	145.5
Panama	555	2,490	2.5	35%	5%	32
Papua New Guinea	820	5,152	3.9	40%	3%	8.6
Paraguay	3,920	0	4.2	40%	4%	10.8
Peru	6,940	2,414	22.9	38%	4%	17.1
Philippines	0	36,289	64.6	39%	4%	209.6
Poland	3,114	491	38.5	25%	10%	122.3
Portugal	1,214	1,793	9.8	21%	13%	106.6
Qatar	60	563	0.5	28%	1%	45.2
Romania	2,508	225	23.2	23%	11%	96.7
Russia	20,139	37,653	149.0	23%	11%	8.7

Sources: Col 1: CIA, World Factbook 1994; **Cols 2 and 3:** Population Concern, 1993 World Population Data Sheet; **Cols 4 and 5:** World Bank, Social Indicators of Development 1994; **Col 6:** World Bank 1995; **Col 7:** Krishna Ahooja-Patel, 1993. The index is based on seven indicators: income,

5 number of people per hospital bed 1994	6 Purchasing power per person 1993 U.S. dollars	7 Women's advancement 1993 index (1 = best)	8 Number of cars 1993 thousands	9 Number of letters and cards posted per person 1992	
399	290	62	17.0	0.18	Laos
72	2,030	–	411.7	23.60	Latvia
230	–	59	179.3	–	Lebanon
597	660	39	–	–	Lesotho
607	–	74	10.0	–	Liberia
246	–	80	800.0	–	Libya
79	1,310	–	657.4	11.00	Lithuania
84	35,850	–	214.4	315.55	Luxembourg
–	780	–	–	–	Macedonia
1,140	240	75	85.0	1.99	Madagascar
645	220	94	25.0	–	Malawi
430	3,160	27	430.7	38.73	Malaysia
1,400	300	103	26.0	–	Mali
100	–	–	–	78.51	Malta
–	–	–	–	–	Marshall Islands
1,278	510	101	11.5	–	Mauritania
300	2,980	37	73.6	19.55	Mauritius
801	3,750	32	10,398.7	6.29	Mexico
77	1,180	–	242.3	–	Moldova
87	400	–	46.9	–	Mongolia
309	1,030	79	775.0	6.84	Morocco
1,280	80	63	45.0	–	Mozambique
1,591	–	–	65.0	1.51	Myanmar
–	1,660	–	109.9	41.20	Namibia
4,010	160	93	–	1.49	Nepal
170	20,710	11	6,303.2	–	Netherlands
149	12,900	–	1,902.8	–	New Zealand
538	360	33	64.0	–	Nicaragua
1,974	270	95	35.0	0.46	Niger
599	310	89	1,350.0	3.09	Nigeria
211	26,340	1	1,961.1	128.16	Norway
558	5,600	67	195.0	–	Oman
1,769	430	90	1,027.4	5.61	Pakistan
300	2,580	25	223.6	1.61	Panama
299	1,120	73	39.0	–	Papua New Guinea
1,087	1,500	60	94.0	0.68	Paraguay
708	1,490	53	671.5	0.55	Peru
780	830	35	1,290.0	13.89	Philippines
153	2,270	–	7,889.9	25.04	Poland
816	7,890	–	2,715.7	65.64	Portugal
340	15,140	–	170.0	–	Qatar
113	1,120	–	1,971.7	11.32	Romania
73	2,350	–	27,500.0	19.67	Russia

maternal mortality, infant mortality, life expectancy, post-secondary school enrolment, share of paid employment and participation in national parliament;
Col 8: Motor Industry of Great Britain, *World Resources 1994-95*; **Col 9:** Union Postale Universelle, *Statistique des services postaux 1992*.

	1 International frontiers 1994 kms		2 Population mid-1993 millions	3 Young and old people 1994 percentage of population		4 Population density 1993
	land	coastline		under 15 years	over 65 years	people per km
Rwanda	893	0	7.4	49%	2%	271.1
Saudi Arabia	4,415	2,640	17.5	39%	2%	7.6
Senegal	2,640	531	7.9	47%	3%	38.8
Seychelles	0	491	0.1	35%	6%	245
Sierra Leone	958	402	4.5	44%	3%	59.1
Singapore	0	193	2.8	23%	6%	4,456.5
Slovak Republic	1,355	0	5.3	25%	10%	49.04
Slovenia	1,045	32	2.0	21%	11%	98.9
Solomon Islands	0	5,313	0.3	47%	3%	11.3
Somalia	2,366	3025	9.5	46%	3%	12.6
South Africa	4,750	2,798	39.0	40%	4%	31.8
Spain	1,903	4,964	39.1	19%	14%	77.3
Sri Lanka	0	1,340	17.8	35%	4%	262.9
Sudan	7,687	853	27.4	46%	2%	10.3
Suriname	1,707	386	0.4	41%	4%	2.5
Swaziland	535	0	0.8	47%	3%	47.6
Sweden	2,205	3,218	8.7	19%	18%	19.2
Switzerland	1,852	0	7.0	17%	15%	165.2
Syria	2,253	193	13.5	49%	4%	67.7
Tajikistan	3,657	0	5.7	43%	4%	38
Taiwan	0	1,448	20.9	26%	7%	–
Tanzania	3,402	1,424	27.8	47%	3%	26.7
Thailand	4,863	3,219	57.2	29%	5%	111.4
Togo	1,647	56	4.1	49%	2%	66.4
Trinidad & Tobago	0	362	1.3	31%	6%	244.3
Tunisia	1,424	1,148	8.6	37%	5%	50.4
Turkey	2,627	7,200	60.7	35%	4%	73.6
Turkmenistan	3,736	1,768	4.0	41%	4%	7.7
Uganda	2,698	0	18.1	49%	3%	71.6
Ukraine	4,558	2,782	51.9	22%	12%	86.2
United Arab Emirates	867	1,318	2.1	35%	1%	19.6
United Kingdom	360	12,429	58.0	19%	16%	235.4
United States	12,248	19,924	258.3	22%	13%	27
Uruguay	1,564	660	3.2	26%	12%	17.5
Uzbekistan	6,221	0	21.7	41%	4%	46.8
Venezuela	4,993	2,800	20.7	37%	4%	21.7
Vietnam	3,818	3,444	71.8	39%	5%	204.3
Western Sahara	–	–	0.2	–	–	–
Yemen	1,746	1,906	11.3	49%	3%	2.1
Yugoslavia	2,246	199	9.8	23%	11%	418.4
Zaire	10,271	37	41.2	43%	3%	16.5
Zambia	5,664	0	8.6	49%	2%	10.7
Zimbabwe	3,066	0	10.7	45%	3%	25.7

Sources: Col 1: *CIA, World Factbook 1994;* **Cols 2 and 3:** Population Concern, *1993 World Population Data Sheet;* **Cols 4 and 5:** World Bank, *Social Indicators of Development 1994;* **Col 6:** *World Bank Atlas 1995;* **Col 7:** Krishna Ahooja-Patel, 1993. The index is based on seven indicators: income,

5 mber of people per hospital bed 1994	6 Purchasing power per person 1993 U.S. dollars	7 Women's advancement 1993 Index (1= best)	8 Number of cars 1993 thousands	9 Number of letters and cards posted per person 1992	
605	200	71	25.0	–	Rwanda
401	7,780	54	2,675.0	21.54	Saudi Arabia
1,385	730	69	115.0	0.80	Senegal
201	6,370	–	–	40.61	Seychelles
830	140	100	47.7	0.24	Sierra Leone
275	19,310	21	435.0	108.25	Singapore
–	1,900	–	–	–	Slovak Republic
–	6,310	–	640.8	–	Slovenia
179	750	–	–	–	Solomon Islands
1,333	–	96	15.0	–	Somalia
–	2,900	43	5,108.3	–	South Africa
209	13,650	19	15,799.1	–	Spain
365	600	40	330.0	21.56	Sri Lanka
959	–	92	66.7	0.18	Sudan
112	1,210	–	–	–	Suriname
296	1,050	–	38.0	8.75	Swaziland
161	24,830	4	3,905.6	–	Sweden
93	36,410	8	3,409.1	–	Switzerland
920	–	50	360.0	0.84	Syria
96	470	–	13.8	–	Tajikistan
–	100	–	3,622.5	–	Taiwan
938	2,040	45	85.0	3.20	Tanzania
620	330	–	3,016.5	11.25	Thailand
686	1,610	82	37.5	–	Togo
201	3,730	22	203.6	10.95	Trinidad & Tobago
516	1,780	46	575.0	12.38	Tunisia
505	2,120	–	2,915.9	17.44	Turkey
92	1,380	–	–	–	Turkmenistan
1,248	190	83	43.1	0.55	Uganda
7	1,910	–	–	5.13	Ukraine
346	22,470	41	335.0	29.99	United Arab Emirates
160	17,970	9	26,838.7	298.12	United Kingdom
194	24,750	7	189,694.0	360.01	United States
221	3,910	29	425.0	2.92	Uruguay
83	960	–	–	–	Uzbekistan
385	2,840	–	1,981.8	3.62	Venezuela
261	170	30	124.5	–	Vietnam
–	–	–	–	–	Western Sahara
1,136	–	105	700.0	0.27	Yemen
–	–	–	3,700.0	–	Yugoslavia
701	–	85	180.0	–	Zaire
284	370	86	165.0	–	Zambia
1,959	540	56	225.0	–	Zimbabwe

maternal mortality, infant mortality, life expectancy, post-secondary school enrolment, share of paid employment and participation in national parliament;
Col 8: Motor Industry of Great Britain, *World Resources 1994-95;* **Col 9:** Union Postale Universelle, *Statistique des services postaux 1992.*

NOTES

1 EXIT

'Human disturbance' is a measure of the threat to natural habitats and the life that depends on them. The World Resources Institute classifies human disturbance as high, medium or low. This map concentrates on what is defined as high human disturbance. But it should be interpreted with caution. Djibouti, for instance, does not appear on the map because both the high and low human disturbance in this state is zero; however, the medium human disturbance is 100 percent. These measures are given only for the proportion of the land area affected. Human disturbance at sea and along the coastline in particular can have a devastating effect directly or through the food chain. This is not recorded in the available sources.

The extinction of certain mammal and bird species is more than impoverishing in itself. It serves as an intimation of what may threaten all life on our planet, in the reckless pursuit of what is regarded as progress. Mammals and birds are not the only life forms with species in imminent danger of extinction. Many species of plants are disappearing, too.

2 MAYHEM

Some states, generally the rich ones, are taking measures to expand their forest cover, since trees take in carbon dioxide and give out oxygen in a natural process that serves to reduce the impact of the 'greenhouse' gases emitted elsewhere. Reforestation does not, however, prevent the same states from engaging in a timber trade which induces other states, generally the poor ones, to continue denuding their own forest cover. The same hypocrisy laments the growth in air pollution whilst at the same time encouraging the production and sale of fossil fuel driven vehicles which are a major cause of such pollution.

The evidence and implications of global warming are increasingly clear, not least to the insurance business, which has its own direct interest. In 1994, Frank Nutter, President of the Reinsurance Association of America, warned: 'The insurance business is first in line to be affected by climate change....it could bankrupt the industry.' But even long-complacent governments are becoming alarmed. A secret UK government report of 1990 revealed this in a warning that parts of the country would have to be abandoned in the event of sea-level rises.

Other forms of environmental mayhem are illustrated by major chemical incidents or accidents cited by the United Nations Environment Programme. These demonstrate an alarming readiness, in the interest of private profit or the supposed growth of the national economy, to take risks not just with the safety of particular communities but with the biosphere itself.

3 THE SCORCHERS

In the wake of the warmest November of the warmest year (1994) ever recorded, it is difficult to deny confidently that the world is heating up, particularly after our experience of the 1980s as the warmest decade ever. It is also difficult to reject the forecast by the United Nations Intergovernmental Panel on Climate Change, that – whatever measures we may take now – average global temperatures will have risen in 2030 by about 2°C above pre-industrial levels (1.1°C above 1995 levels) and by 2090 by 4°C (3.3°C above 1995. This would make the earth hotter than it has been for 120,000 years.

There is a strong connection between global warming and the buildup of greenhouse gases resulting from human activity. These gases are primarily made up of carbon dioxide arising from the use of fossil fuels, cement-making and deforestation; chlorofluorocarbons (which are mainly responsible for the depletion of the ozone layer) arising from refrigeration, the manufacture of solvents and foam products; and methane, from wet rice cultivation, factory farming and the production and transport of natural gas. Together these account for 86 percent of the human-generated greenhouse gases.

We are doing very little to restrain the buildup of greenhouse gases. Energy efficiency varies widely from country to country and, within countries, between users. International agreements to limit emissions are easier to sign than to enforce.Even an agreement on what to do is as difficult as the doing.

4 THE GUZZLERS

Energy is the feedstock of material abundance and personal choice. Hundreds of millions of people will suffer hunger, disease, discomfort and physical and mental confinement because they do not have access to enough of it. But if we continue to produce and consume energy in the quantities and in the ways we do now, even more people will suffer, perhaps terminally.

The dilemma might be avoided in part by using energy less prodigally. Just five percent as much energy is needed to recycle aluminium as to produce it from bauxite, the original raw material. For steel produced entirely from scrap, the savings amount roughly to two thirds. Newsprint from recycled paper takes 25-60 percent less energy to make than that from wood pulp. Recycling glass saves up to a third of the energy embodied in the original product.

The joule is a general measure of energy use, equivalent to 0.239 calories. One gigajoule equals one billion joules and is equivalent to 34.1 tonnes of coal or 23.8 tonnes of oil.

5 INFERNAL COMBUSTION

How good and yet how bad is the car. It is a device that enables us to move further and carry more than ever before, but hinders our access to where we want to be; that saves time, but dissipates it by extending the distances we travel; that covers ground, and buries it under tar or concrete; that enlarges personal freedom, and undermines the social connectedness that gives it meaning; that clears cities of refuse, and pollutes their atmosphere; that saves lives and maims them.

Directly and indirectly, the car is the most expensive means of transport in use: in air pollution costs, it is sixteen times more expensive than trains, twice as much as planes; in CO^2 emissions, the car costs twice as much as trains (although half that of planes); in noise pollution, it is four times as costly as trains (although less by a quarter as noisy as planes). Cars are respectively ten times and 69 times more costly in accidents than trains and planes. Altogether, the external costs of motor passenger transport are 725 percent higher than transport by train, and 190 percent higher than transport by air.

This does not stop governments from continuing to encourage car ownership and to invest in road building rather than in public transport systems.

6 URBAN BLIGHT

Shrinking economic opportunities in rural areas and the lure of enhanced opportunities and services in cities combine to promote a process of urbanization that is all but ubiquitous. Furthermore, the process is accelerating. From 1875 to 1900, the annual rate of urban growth in the rich world was 2.8 percent. In the poor world, urban populations grew at an annual rate of around 4 percent from 1975 to 1990. On present trends, half the world's population will be urbanized by the year 2005, and two thirds by the year 2025. By the year 2025, some 4 billion people in developing countries will be classified as urban, equivalent to the world's total population in 1975.

For all the advantages in employment, education and access to medical care that may benefit many people in urban areas, the environmental impact of the megacities is likely to be formidable. City residents tend to consume more industrial goods and energy-intensive services, especially in developing countries, and urban populations everywhere create concentrated air pollution, water pollution and solid waste.

7 MULTIPLICATION

China's 'floating population' of illegal city dwellers numbers 100 million or more; and the population of hidden 'black babies', born despite the official one child family policy, must be counted in the tens of millions. Elsewhere population censuses are the outcome of politics rather than their shapers; in Nigeria or Lebanon, for example, where censuses are routinely rigged or avoided, and even in the stable, rich countries whenever constituencies are adjusted to take account of demographic changes. There are isolated, marginal people everywhere: hostiles, illegals, anti-socials and a-socials, who have cause to avoid officialdom. And there are innumerate census enumerators, often engaged in more pressing business, who prefer the cool of their office to the heat and possible danger of the world outside.

If it were not for the international agencies, such as the Population Division of the United Nations, we would hardly know how many we are, where we are, and how fast our number is growing. They provide annual estimates of all states' populations and an essential critique of the often self-serving national figures. However, even UN figures, particularly its forecasts, should be treated with caution. Reproduction habits can change unexpectedly; wars can deplete and disperse whole populations; mass migrations can swarm suddenly. At the end of the 1980s, no fewer than six states – Afghanistan, Angola, Cambodia, Chad, Laos and Nicaragua – were thought to have less than 90 percent of the population that had been predicted for them at the beginning of that decade.

The population projections in the map are based on reasonable assumptions about the future course of fertility, mortality and migration. They do not allow for sudden eruptions of war, natural disasters or unexpected changes in disease patterns.

States with under 0.05 percent of the world's population – less than 2.75 million today and 4.21 in 2025 – do not feature in the cartogram. There are 53 of these, and their combined population in mid-1993 was 47 million.

8 DIASPORAS

People do not always stay in the countries or even the continents where they were born. Some go, or are taken, in large numbers elsewhere in the world. These are the origins of diasporas.

As widely differing estimates for the black components in the populations of Brazil, Cuba and Colombia indicate, the extent of the black diaspora is nowhere more difficult to establish than in an agreed classification of who is black. Despite colour distinctions within the black community of the U.S.A., blacks are defined there, not least by themselves, as people of part or whole African ancestry. In Brazil, where people define themselves, for the purposes of the census, as white, yellow, red, brown or black, the bias is strongly towards levitation as far up the scale of prevailing racial value as credibility suggests, and often further. In Cuba, where the government has long regarded racial identity as a domestic distraction when not a political deviation, official census returns have remained undisclosed and are likely to be unreliable. We have chosen to use the U.S. demographic definition and to rely on such expert estimates as are available. In general, the higher estimates are likely to be the more valid ones.

We have not included in the map on the black diaspora every country outside Africa where any blacks live, since this would scarcely be of much relevance or value. Where the black component of the population is below one percent, we have identified only those countries with still substantial communities or where there is, however small the communities, some evidence of related social tension. Blacks are all too often a prime target of white racism, and discrimination against them takes numerous forms. We have given examples of functional segregation in the U.S.A., but it is widespread in other countries, such as the UK, too. Despite traditional and official pretences to the contrary, for instance, Brazilian society is profoundly racist, but the information on functional segregation is not available to illustrate it here.

The Indian diaspora also provides difficulties of demographic definition, though for different reasons. The India of the British Raj, from which so many indentured labourers were drawn, became with independence the states of India and Pakistan, and Pakistan subsequently became the two states of Pakistan and Bangladesh. For most countries, we have defined Indians as members of communities which originated from the Indian sub-continent, since separation of them into Indians, Pakistanis and Bangladeshis is not possible and would make no historical sense. Where official statistics make the relevant distinction, as in the UK where such settlement is overwhelmingly recent, we use only those for Indians. For Pakistan and Bangladesh themselves, we use the definition of Hindus.

The Chinese diaspora produces no such difficulties. It has been chosen, rather than the Armenian or Greek diasporas, for instance, since it is, like the black and Indian diasporas, a target of hostility and discrimination in certain societies.

9 THE QUANTITY OF LIFE

Living long and living well are not the same. This map concentrates on physical survival and the conditions that make it possible, primarily safe water and adequate health care. What it feels like to be alive is shown in the following map (The Quality of Life).

The indicators we have for both maps are horribly crude. Take health care, for instance. It can be safely assumed that almost everyone in the U.S.A. could, if need be, reach a medical facility within an hour using normal means of transport – the conventional criterion of access to care. Yet there are between 31 and 37 million Americans who are not covered by medical insurance, and many more who are underinsured, and so effectively denied medical attention. In theory, all Russian citizens are covered by medical insurance; in practice, some of the treatment on offer is so poor and the facilities so unhygienic that they are avoided at all costs.

Over-aggregation (the loss of relevant detail in making broad categories) affects this map, and the one following, in other ways. Life expectancy at birth – the basic ingredient of the main map – takes no account of the fact that women generally live significantly longer than men. The data on access to safe water conceals the enormous differences in provision between rural and urban areas in a bland, often meaningless, national average.

Over-aggregation is not the only, or indeed, the most grievous of the data's sins. Simply, they contain too many round fractions – halves, thirds, quarters – to inspire confidence. They reek of the office and of written records, not of the real world with all its particularities. But they are the best we have and approximate reality even if they do not represent it accurately.

10 THE QUALITY OF LIFE

If people were allowed to move about freely, we might learn a great deal about how they experience life in different places. Since they are not, we have to make do with conjecture and with measurements that rest on the wobbliest of assumptions, namely, that what makes for happiness – or misery, as the case may be – is more or less the same everywhere; that people can increase the one, or avoid the other, through personal commitment and political will; that meaningful comparisons can be made between one person and another, and between one society and another.

These assumptions reveal more about the measurer than the measured. They reflect a view of people as social, mutually supportive creatures and a view of world society as essentially a unity marred occasionally by conflict, not a disparity occasionally drawn together in harmony. It is a view held by a relatively small circle of privileged, Western-educated intellectuals steeped in world affairs – a humane, optimistic view with which, broadly, this atlas agrees. But it is not a majority view, and the indicators of human fulfilment or development that it propounds are not objective.

The Human Development Index compiled by the United Nations Development Programme is the most ambitious of its type. It combines life expectancy at birth (between 25 and 85 years), adult literacy (0 to 100 percent), years of schooling (between 0 and 15) and real income (between 200 and 40,000 'international' dollars of equal purchasing power).

Some of the indicators it uses are weak because they rest on statistical sand; some because of the nature of the evidence; some because of the way that evidence is gathered. All suffer because they rest on national averages, as if there were, or could be, an average human experience.

The inset map and graphic illustrate harder evidence for the stresses of life in the contemporary world: the needless shortening of its span, despite the miracles of modern medicine, and its undervaluation.

11 RICH AND POOR

1.3 billion people cannot afford a nutritionally-adequate diet and essential non-food requirements; they live in absolute poverty. On the other hand, there are 223 billionaire families who live in absolute riches, scarcely able to consume personally the interest on the interest from their investments.

Of the 53 countries for which comparable estimates exist, Botswana and Brazil, half way down the list in average income per head (23rd and 28th respectively), nonetheless top the bill in income inequality; Sweden, Norway and the Netherlands (7th, 11th and 15th respectively in average income per head) show as much inequality as Ethiopia or Uganda, the two poorest in average income per head.

12 BREADLINES

The availability of adequate nourishment in a particular state is not the same as adequate provision for all of its inhabitants. Nor do calories themselves guarantee health. Other constituents of a diet, such as adequate protein and vitamins, are necessary. To take only one example, Mexico has available 131 percent of the average calorie need, yet 14 percent of its children under the age of five are estimated to suffer from malnutrition, and the growth of 22 percent of those aged from 24 to 59 months is stunted.

Statistics for available average calories are duly listed for the rich countries, which do not provide estimates for the incidence of malnutrition and stunting among children. Such absences from the statistics do not necessarily match reality. In some rich countries, such as the U.S.A. and UK, the incidence of poverty and deprivation is not only widespread but rising.

13 CONNECTIONS

In 1965, 85 percent of the world's 150 million telephone lines were in Europe and North America. Fax was almost unknown and computer networking a contradiction in terms. There was just one trans-Atlantic telephone cable which could handle 89 calls at a time, and a handful of radio links across the Pacific. 'Mobile' telephones were unknown. A three minute phone call from the U.S.A. to Europe or Asia (assuming you could get through) cost, typically, U.S. $90 at 1995 prices (U.S. $18 at the time) and much more if the call were made in the other direction.

In 1995 the global network comprised more than 600 million telephone lines and over 1.2 billion terminals in some 190 countries. There were faxes galore, computer networks travelling into information highways, and ever proliferating 'mobiles'. The cable and satellite network across the Atlantic handled almost a million simultaneous calls at between U.S. $1 and $1.50 per minute from Europe, and half that price from the U.S.A. Satellite communications now cover the world.

You might say that a great deal has changed in one generation. But not that much – as yet. CNN, the satellite-distributed news network, and Internet, the transnational matrix of computer networks, together reach no more than two percent of the world's population. The rest are dependent on 'snail mail' – the postal system – as the mainstay of their contacts beyond the narrow horizons of face-to-face, local communication and personal travel. In 1991 the average amount of time spent on international calls per year – about seven minutes – straddled a huge range, from seven hours per citizen in Luxembourg to six seconds in Africa.

14 ILLITERACY

Many states neglect to record the problem of illiteracy in their midst and others underestimate the number of illiterate adults in the population. For this map, therefore, we have had to rely on estimates, supplied by UNESCO and the CIA. UNESCO records nothing more precise than an illiteracy rate of below five percent in advanced industrial states.

Symbols on the map illustrate the ratio of female to male illiteracy. Most of the recent statistics available demonstrate that the difference between the proportions of illiterate men and women is marked and sometimes enormous. However, recent statistics are simply not available for many of the industrialized states. Only Greece, Italy, Portugal and Spain provide relevant statistics for as late as 1990. Those provided for Belgium, France and the U.S.A., as for Bulgaria, Hungary and Romania, date back to 1970. There are no relevant statistics available for the successor states of either the USSR or Yugoslavia, and the statistics we have used belong to these states before their break-up. Those for Yugoslavia, at least, date back to only 1990. Those for the USSR date back to 1970.

15 WASTING LIVES

The high rates of mortality among the children of the poor represent a hideous and unpardonable waste of human life. Nor is that by any means the only such waste. For many, mere survival is, in another sense, a deadend. Children who survive are often deprived of the education that might otherwise enable them to realize their potential. As adults, they are often relegated to the edges of the economy, where their lives are at risk from their inability to earn an adequate wage.

16 MISSING WOMEN

By law or convention, all but a very few women are denied control over the productive assets that lend economic independence. In some (poor and not so poor) countries where land is particularly important in sustaining life they are denied ownership rights to it, or equal ownership rights with men. In others, even some of the richest states, they are denied finance through mortgage-lending or banking policies which are defined in a way that assumes men to be the dominant economic agents and thus perpetuates the status quo. In Iran, for instance, a widow inherits a mere eighth of her husband's estate; a daughter gets one share for every two inherited by her brother. From that primordial denial flows a secondary denial – of education – and a tertiary one – of influential, well-paid work.

There is a substantial body of data on women's share of education, of certain jobs, of the market in consuming this or that, but not of women's share in the ownership and control of productive assets. Nobody really knows much about discrimination in access to productive property. There are, as yet, no universally agreed indicators, let alone any serious attempts to measure its incidence worldwide. The United Nations Development Programme promises to address this gap in our knowledge in its Human Development Report for 1995. Meanwhile, and long after that is published, women will continue to work twice as hard as men for half or less in paid income; will continue to have unwanted babies, usually the only dependable source of additional labour and income (in poor countries), and of status, pride and spiritual nourishment (in all countries).

The cumulative result of discrimination and social impotence is that there are far fewer women than there would be in a more even-handed world. In some countries, such as China, many female babies are simply killed at birth; in others, such as India, many are overworked, undernourished and neglected to destruction. The number of 'missing women' is the difference between the number of females in a population and the number who would be alive if they retained their natural female to male ratio at birth (94.6:100).

17 BODY POLITICS

The battle lines between the private, creative and caring realm of parenting and the public, impersonal, instrumental world of population policies and state interest run through the very centre of women's bodies. The outcome of that battle is in constant flux: sometimes, when labour is scarce as in the rich states in the 1960s, women find it possible, although never easy, to extend the frontiers of control over their bodies. In times of recession and idleness, women find such frontiers contracting. Nowhere is this more apparent than in Eastern Europe where the collapse of Communism, profligate with labour as with so many other things, has meant a sharp turnabout in what were once among the most liberal abortion regimes in the world. The personal has paid heavily for the political in those countries.

The law on abortion does not everywhere reflect clinical practice. In most Muslim states, and in many Latin American and African ones, the law is more liberal than the reality; while in Israel, New Zealand and South Africa, legal abortion rates approach those in states which permit abortion on demand. Interpretations of the same law can vary widely within a state, as in Switzerland. In some ostensibly restrictive states – Bangladesh for example – menstrual regulation, that is, early abortion without a pregnancy test, is freely available; whereas in some ostensibly liberal states – Togo, for example – legal abortions are few in number because the services are scarce or not available.

The worldwide trend towards liberalism in abortion law since the 1960s is, for these reasons, more a matter of ideology and political rhetoric than of women's experience. It might well be that the institutionalization of medicine and its embrace of pregnancy and childbirth, have widened women's choices in theory while narrowing them in practice.

Abortion is one method of exercising control over childbirth. Less extreme, although almost as politically charged, is the accessibility of birth control and the range of choices available.

Domestic violence is often unacknowledged and frequently concealed. The victims as well as the perpetrators may go to great lengths to hide evidence of wife-beating and marital rape, so records are haphazard and scarcely scientific.

18 CHILD EXPLOITATION

The most widespread form of abusing children is almost certainly their employment, sometimes below the age of six, as workers. Yet here, as with other forms of child abuse, information is scarce and often merely indicative. An ILO (International Labour Office) study has been undertaken, but information on a national basis is not yet available. The statutory provisions for minimum age requirements, themselves all too often more honoured in the breach than the observance, nonetheless say much of the way children are treated in numerous states. The currently most detailed source is the U.S. State Department's Country Reports on Human Rights Practices. Particular studies of child labour, most disturbing in their details, come from the International Confederation of Free Trade Unions.

The 1989 United Nations' Convention on the Rights of the Child sets out to promote the basic health care and education of the young, as well as their protection from abuse, exploitation or neglect, at home, at work, and in armed conflicts. A recent report on 'The Progress of Nations', from UNICEF (United Nations Children's Fund), testifies to the readiness of governments to sign the Convention, without a commensurate readiness to ratify their adherence or comply with their undertakings. The sexual exploitation of children by foreigners, evidently condoned or inadequately suppressed by most governments, is the subject of one study in the UNICEF report; the incidence of infant deaths from presumed abuse, which highlights the difference between the U.S.A. and Denmark at one extreme and Italy and Spain at the other, the subject of another. In each case, the information covers only selected states, and others are not to be presumed innocent because they are not included.

19 THE KILLING WEED

Half a million Americans are hooked on heroin, 3 million are regular users of cocaine and crack, and 15 million are alcoholics or alcohol-dependants. This is the official count in 1995. Unofficial estimates run three times higher. Taking the unofficial count and adding to it the untold addicts of all the other natural and artificial mood modifiers would still not bring us anywhere near the number of nicotine addicts – perhaps 53 million – spending U.S. $60 billion a year, including taxes, on their habit and probably costing as much again in health care and accidental damage. By contrast, illicit drug use of all sorts is estimated to cost some U.S. $30 billion a year.

The comparison between nicotine and other drugs is as dramatic elsewhere. In the Netherlands, cigarette smoking is reckoned to cause nine times as many deaths as all other addictions. In the UK more than twice the number of deaths are attributable to cigarette-smoking as to self-administered barbiturate poisoning plus alcohol.

Consumers are locked into tobacco's addictive qualities, business into its dependable cash flow, government into its taxability. Nicotine addiction is a public addiction and, for that reason, appears – wrongly – less harmful than its more private, often illegal, competitors. It is a domesticated addiction, not generally linked to crime or violence, so that it appears – wrongly – more acceptable socially. Like all addictions, it appears to offer a sanctuary from stress and failure. It does not.

In the U.S.A. the proportion of adults smoking has dropped from 42 percent to 26 percent since 1955. Almost 90 percent of Americans – including a third of all smokers – tell pollsters they find cigarette smoke annoying. The number supporting a total ban on smoking indoors, as opposed to milder measures such as the installation of no-smoking areas, has nearly doubled since 1983, to 35 percent. Since 1986 the proportion of firms that has banned smoking on their premises has risen from 2 percent to 34 percent. In 1993, the Environmental Protection Agency officially classified secondhand smoke as a health hazard.

With sales of U.S. $45 billion a year, the American tobacco giants are still profitable and powerful. But they are less confident than they have been in the past. For although their sales in Asia and Eastern Europe are expected to explode, the new markets are unlikely to be as profitable in the short run as the domestic market. And in the long run they might well catch the American virus of health consciousness and care.

20 PLAGUES NEW AND RENEWED

The spread of AIDS has been gathering pace, not least in countries where the rates were previously low. In its report of July 1994 on 'The Current Global Situation of the HIV/AIDS Pandemic', the World Health Organisation (WHO) declared: 'Allowing for under-diagnosis, incomplete reporting, and reporting delay, and based on the available data on HIV infections around the world, it is estimated that around 4 million AIDS cases in adults and children have occurred worldwide since the pandemic began'. This represents a 60 percent increase over the estimated 2.5 million cases as of July 1993 or only one year before. The map gives national totals of reported AIDS cases, where these exceed 10,000, up to the end of March 1994. Corresponding totals for the end of 1990, with the percentage increase from then up to March 1994, reveal how rapid and sometimes explosive the rate of increase has been: Brazil 20,578 (140%); Côte d'Ivoire 6,898 (171%); Kenya 15,252 (948%); Malawi 14,851 (115%); Mexico 5,905 (211%); Rwanda 4,489 (138%); Spain 7,304 (231%); Tanzania 22,081 (75%); Uganda 19,955 (120%); U.S.A. 190,774 (116%); Zaire 16,147 (41%); Zambia 4,202 (608%); and Zimbabwe 5,994 (366%).

Tuberculosis was supposed to be in retreat from improving social conditions and the development of new drugs by advances in medicine. Yet, while the incidence may be declining in some states today, it is rising – often sharply – in others. As with diphtheria, another disease emerging as a serious scourge again, this is related to deteriorating social conditions, in poor countries or among the poor within rich ones. In March 1995, the World Health Organisation predicted that tuberculosis will kill at least 30 million people over the next decade and could infect some two billion, or more than a third of the world's population. In particular, it warned that unless remedial action was taken within three years, tuberculosis would assume catastrophic proportions in many of the successor states of the former USSR.

21 WANTED HANDS

The labour force is conventionally divided among three categories of economic activity. Agriculture includes hunting, fishing and forestry as well as farming. Industry includes mining, construction, water, gas and electricity, as well as manufacturing. Services account for the rest, such as wholesale and retail trading; restaurants and hotels; transport and communications; finance, insurance, real estate and other business services, as well as social and personal ones.

A high proportion of employment in agriculture is generally an indication of economic backwardness and poverty. A high proportion in industry, however, is not a sure sign of corresponding economic advancement. Some of the highest proportions are in Eastern European states and successor states of the former USSR where productivity is low, and what is produced is not always marketable. A high proportion of employment in services is often taken as a more reliable guide to economic advancement, but the correlation is not invariable. At one extreme, such employment may reflect a substantial sector of involvement in finance and insurance, information technology and telecommunications; at another it may merely reflect a substantial dependence on tourism or a disproportionate government bureaucracy.

22 VALUE SUBTRACTED

The term 'exploitation' has largely disappeared from the vocabulary of current politics. This does not, however, make the fact of exploitation any less real. And whatever it may be called, its existence is implicitly recognized by the measure of 'earnings as percentage of value added' (value added being what is charged for the product after deduction of material costs) in the World Bank's information on 'manufacturing activity'.

The rates of exploitation which the paucity of relevant statistics allows do not coincide with the differentials between rich and poor countries. Niger is at one extreme and the Central African Republic at the other, for instance. In both cases, the size of the manufacturing sector in relation to the economy is so small that quirks may be responsible for the differential. Regrettably there are no relevant statistics for the successor states of the former USSR. Yet, for all of the reservations we have about the data on which the map is based, the information it contains remains useful, and illustrative of the phenomenon with which it deals.

23 THE LABOUR FORCE

The right of workers to combine in advancing their own interests against the power and exactions of their employers was dearly bought and was a crucial contribution to the advance of democracy itself. It is a right that lost none of its meaning for being abused, as it was in the former USSR, and other one-party states, which made trade unionism an important component in securing the obedience of labour and its service to the regime. Even in the traditional democracies trade unions have not always been innocent of abusing democracy itself, as officials have gained or perpetuated control by manipulating election.

Yet the right remains more important than its abuse, and has in recent years been notably asserted by trade unions in the cause of democratizing their societies. Trade unions were central, for instance, to the defeat of the authoritarian regime in Poland, as in other states of the former Soviet system. In Latin America, especially in Argentina, Bolivia, Brazil, Chile, Ecuador, Peru and Uruguay, unions were crucial in the popular and eventually successful challenge to repressive regimes. The same has been true in Africa. Revolt against one party regimes in Congo, Mali, Niger and Zambia has been led by trade unions. Even as these notes are being written, trade unions in Nigeria are challenging military rule. In other states, trade unions are asserting their independence.

It is scarcely surprising that in a number of countries where monarchical, theocratic or other forms of unelected rule do not encumber themselves with even the pretensions to democracy, trade unions are banned altogether or allowed to exist only in conditions of the closest government control. Yet trade unions are also under attack in parts of the avowedly democratic and industrially advanced world. Recent legislation in the UK, for example, on the pretext of dealing with trade union abuses, has abridged their rights as well, and membership has declined steeply. There, and in other states, this decline in trade union membership is partly due to the economic shift from industry, the traditional home of trade unionism, to the service sector, but it is also due to a rising rate of unemployment, which has encouraged employers to be more exacting in their requirements. Employers become more determined to hire only non-unionized labour, and workers more willing to accept this condition. There is a clear connection, for instance, between the decline of union membership in the U.S.A., from 30 percent of the workforce in the mid-1960s to 16 percent in the mid-1990s, and the fact that in 1991 unionized workers there were paid on average 23 percent more than those in non-unionized enterprises.

In other democratic states, however, and in those which are enjoying industrial success, such as South Korea, trade union membership has been rising rapidly.

Statistics on strikes and lock-outs are available only for a relatively few countries. We have included such information, however, to illustrate continuing labour militancy and the varying rates of this from one country to another.

24 WOMEN IN WORK

Statistics for the participation of women in the labour force must be treated with caution, especially those for the poorer countries. Women frequently work in the informal sector of the economy, selling food and trading in household goods, or in unregulated industries and services, where their activity goes largely unrecorded. In poorer countries, their work as subsistence labour in agriculture is all too often assigned little or no economic value by governments and ignored in labour statistics. In Pakistan, for instance, such statistics exclude an estimated 12 million agricultural workers. Domestic labour outside the home is similarly disregarded, even though collecting and transporting food and water, for example, are essential activities which may require several hours of work each day.

In theory, the proportion of women in parliament should advance in step with democracy. In South Africa, the first democratic election, in April 1994, raised the proportion from 3 percent to 25 percent. This is an unusual correspondence, however, as is all too clear from the low proportions of women in parliament in such long lived democracies as those of France, the UK and the U.S.A. Real democracy has yet to surmount the gender issue.

25 UNEMPLOYMENT

The officially high rates for unemployment in most of the advanced industrial states, where governments were not so long ago committed to objectives of full employment, are socially wasteful and unsettling enough. But the real rates may well be significantly higher than the recorded figures suggest. The rate of unemployment is a sensitive political issue, and the official figures record only those officially defined as unemployed. In the UK, for instance, such definitions have been frequently revised, and a substantial number of the unemployed may be concealed in other categories such as that for recipients of invalidity benefits. In some advanced industrial states, unemployment is a term applied only to job-seekers, and unemployment benefits are paid only for a stipulated period. In such circumstances, unrecorded numbers of the unemployed may stop seeking jobs after repeated experiences of failure and may simply rely on basic social benefits.

Statistics for the successor states of the USSR are not so much manipulated as irrelevant, since many of the supposedly employed are in state enterprises which function fitfully and pay wages that consign their employees to poverty, if not destitution. Figures for many poor countries are either unavailable or apply only to a relatively small sector of the economy, to the urban areas, or only to the capital city. The figures for the rate of underemployment, applied to those people who work in casual, part-time or seasonal jobs, are more significant even though they are usually no more than estimates. Where there are specific reports and/or estimates, we have identified states with high levels of underemployment but these are by no means the only ones.

In a number of poor countries, figures for unemployment and underemployment would be much greater, were it not that many in the workforce are working abroad. In 1988, for instance, two and a half million Egyptians, equivalent to one sixth of that country's workforce, were estimated to be working abroad, most of them in Saudi Arabia and the Gulf states. At any one time, some 60 percent of the active male labour force in Lesotho are working in South Africa. Relatedly, figures for the net migration rate in 1993 are merely illustrative and partly deceptive. Migrants who leave may be balanced by migrants who return, on visits or at the end of employment contracts. For countries such as Egypt, the limits of labour recruitment abroad may well have been reached, as traditional recipients tighten their immigration policies.

26 A GLOBAL GRASP

The combined foreign sales power of the world's 100 top transnational corporations (TNCs) represent well over 40 percent of the world's total export trade. This is not the full extent of the TNCs' power. They can and often do have a marked impact, through the policy decisions they make, on the economies and politics of states. Except for a small minority in the state sector of the countries where they are domiciled, TNCs are effectively their own masters, answerable only to their shareholders. Since in almost all cases the shareholdings are widely diffused, this amounts to their being answerable only to themselves in the shape of their management boards.

In 1991, almost half (48.4 percent) of all sales by the 100 top TNCs came from countries other than their own, and well over a third (37.8 percent) of their total assets were held abroad. Foreign direct investment by TNCs totalled over U.S. $148 billion, and by far the bulk of such investment – almost U.S. $108 billion – went to the already developed economies. In the main, therefore, the TNCs invest in one another's backyards. While France, for instance, attracted U.S. $15.2 billion and the UK, U.S. $21.5 billion in 1991, China attracted only U.S. $4.4 billion and India a mere U.S. $961 million.

27 HEDGES AND BETS

Investment and speculation are as old as markets themselves. It has long been possible to buy shares on the stock exchange by contracting to pay for them at some stipulated future date or by trading in them on the basis of a 'margin', or backing, through the provision of cash or shares to meet a stipulated proportion – say, ten percent – of the value. It was essentially this 'margin' trading, on the great 'bull' or rising market in the late 1920s, that turned the fall on Wall Street in 1929 into a collapse, as 'margins' disappeared and the shares that provided them were sold to cover commitments. It has also long been possible to speculate similarly in the hope of a 'bear' or falling market, by selling shares without owning them, on the basis that they can be bought more cheaply later with a corresponding profit. The consequence of misjudgement or misfortune can be no less calamitous in this exercise. And what is done with shares can be done also with commodities, from copper and oil to pork bellies and frozen orange juice.

Increasingly, moreover, as financial markets have become more sophisticated, they have developed numerous new instruments, which may be used to hedge against risk but also to increase it. It is now possible to speculate not only on the future value of particular shares and commodities, but on future movements in the value of various currencies, interest rates, one or other stock market index, and in options or rights to undertake such transactions at a stipulated price by a stipulated date. On this basis, for instance, share prices can often be affected less by investment considerations than by speculative positions on the future movement of the stock exchange index.

The growing dominance of these so-called derivatives is being accompanied by growing disquiet among those responsible for financial regulation. The global nature and complexity of the derivative markets involve so many interlocking players that the risk is real of one financial collapse setting off others in an uncontainable series, whatever the underlying state of the real economy. When an oil company in Japan can lose U.S. $1.5 billion in trading foreign exchange derivatives, a German industrial company a similar sum by betting on movements in the oil market, and a U.S. soap and detergent firm over U.S. $100 million on interest rate derivatives, the question is whether the scale of speculation has not begun to dwarf that of investment.

As this book was going to press, Baring Brothers, the blue-blooded British merchant bank, collapsed after losses – estimated at close to £1 billion – in Japanese stock market index derivatives.

28 THE GOLDEN FIX

In October 1994, officials and ministers from more than 120 states gathered in Naples for a United Nations conference on organized international crime. Little came of it, of course. For however sincere individual representatives might have been about wanting to collaborate with one another, they were hampered by a combination of political rivalries, jealously guarded national legal systems, official corruption and sheer incompetence. And, as likely as not, they had been pre-empted by representatives of the new style multinational criminal corporations meeting in a hotel somewhere, in Eastern Europe perhaps.

These criminals are not cast in the mould of Al Capone, a near illiterate who acquired millions and did not know how to dispose of them, who wore flash suits and cashmere coats, who used naked fists and guns hidden in violin cases to get to the top, and who died of syphilis and paresis of the brain. The new hoods are cosmopolitan businessmen, well-educated, well-spoken, who know how to move among politicians and officials and transfer money from Wall Street to London to Paris and beyond. They are privileged people, with plenty of opportunities. They are just more than usually greedy as businessmen and so tempted to take the criminal route.

The profits accruing to them are huge. Pablo Escobar, the drug baron shot dead in late 1993 by the Colombian army with support from the CIA, the U.S. air force and the U.S. drug enforcement agency, is reputed to have authorized payments of U.S. $1 million per day to keep himself out of jail. More must have gone in bribes to keep open trade channels, to launder the proceeds of trade, and in investing in legitimate business. For the aim of the people who direct criminal multinationals is to drop the adjective – every generation has its robber barons eager to become merely barons.

They are willing to pay heavily for it. Members of the Cali cocaine cartel in Colombia, are reputed to pay 17-20 percent for money laundering services; and money brokers are often able to buy cash at a discount of 23 percent. The laundered money finds its way into lifestyle and cash-intensive industries – travel agencies, exchange houses, casinos and other gambling joints, international trading firms – as well as commercial and residential buildings and construction.

It is difficult to paint a dispassionate picture of criminal business. Business generally is secretive; criminal business, however defined (and that is an arbitrary matter) is even more so.

The best available source we have – the U.S. Bureau of International Narcotics Matters – is not without bias. It considers domestic drug consumption and indigenous supply less important than foreign supply and shipment to the U.S.; it values foreign collaboration with the U.S. authorities more highly than foreign action; it paints a better picture of countries favoured for other reasons such as Israel, and reports badly on countries out of favour, such as Cuba.

29 PRIVATIZATION

Privatization is for citizens a second expropriation by the state; the first is the taxation which paid for the assets in the first place.

For the purchasing companies, privatization is a quick, seemingly safe, route to increased profit, presence and power. For the divesting states, the motives are more complex. Some see it primarily as a necessary adjustment to the need for international economic mobility and to greater competition in the global market. Others see in privatization a way to relieve budgetary stress. Yet others use it to keep rich, foreign uncles from meddling in what they consider to be their internal affairs. And others use privatization to spread the charms of possession beyond the political class, and so to widen its base of support.

None of these aims is pursued to the exclusion of the others. Nor are they pursued with clinical purity. The advantages to be gained from breaking up state monopolies are often lost by conniving at or condoning private ones. Attempts to furnish or replenish Treasury coffers are qualified by giveaway prices for the aspiring privilegentsia.

Even assuming that the aim of privatization is clear and uncorrupted, there are natural limits to it everywhere except in the heartlands of the market system. The state has built its business – badly in most cases, madly in some – where the domestic private sector proved unequal to the task of fending off or challenging foreign economic might. The state commandeered enormous resources in the process, further weakening, where it did not extinguish, private enterprise. Now that the major state commercial bodies are seen to be at best awkward players in world markets, or at worst, clear losers, there is usually no one left on whom to unload them except – cruel irony – the foreign private enterprises they were designed to resist in the first place.

In brief, privatization is a primrose path to an unprecedented concentration of economic power in ever-fewer, ever-larger hands on a world scale, and to its equal and opposite – an unprecedented loss of economic power by an ever-larger number.

30 MUFFLED VOICES

The different categories we employ for this map are our own; chosen to define general distinctions among so many particular ones, and to provide some basic historical perspective.

The term 'established multi-party democratic system' implies no comment on the content of the democracy involved. Any category that encompasses Sweden, Senegal and Guatemala is necessarily broad. All that is implied is that a multi-party political system, with elections by supposedly secret ballot on a basis of adult suffrage, has traditionally, if not always continuously, been in place. Where appropriate, certain qualifications have been added, such as interruptions by periods of military rule or the significant incidence of extra-judicial violence as a form of political pressure. The second term, 'recently adopted multi-party democratic system, or in transition to one', identifies states, most notably those that were formerly part of the USSR, which have emerged from a recent past of one-party regimes or, as in the case of South Africa, some other kind of developed repressive rule. The term 'one-party regime' is also broad. It encompasses not only states where this system is constitutionally sustained but also those, such as Mexico and Kenya, where ballot rigging, intimidation, and/or institutional structures of control have secured a one-party regime in fact, if not in form. Terms for the remaining categories require no elaboration.

31 NOTIONAL INCOME

Gross National Product (GNP) and Gross Domestic Product (GDP) are the most widely-used general measures of economic power, and for that reason alone are unavoidable when making international comparisons. They measure the consumption of goods and services that are bought or paid for in taxes. They do not normally count unmarketed goods produced for home consumption, and do not include at all the untraded services, from sex to socialization, which are essential for human life. They therefore underrate the incomes of rural and family-oriented societies and exaggerate those of industrial ones and of societies resting on nuclear and sub-nuclear families. They do not cover the cost, in depletion and damage, of economic activity and, therefore, understate the incomes of traditional and subsistence-agriculture societies where such exist, and overstate those of resource-based, market-oriented economies. They do not distinguish between productive and unproductive activity – between making butter and making guns – and, therefore, confuse economic with military power, in the short term at least. In consequence, whatever their virtues as pointers to the current resources available or potentially available to the state, they are inadequate yardsticks of human welfare or social health.

Even within its own terms, national accounting is flawed. It is insensitive to changes in consumption patterns, and so an imperfect guide to recent changes. Expressed in a single national currency – conventionally, the U.S. dollar – national accounting is at the mercy of exchange rates which can and often do gyrate madly.

A different measure, developed by the UN's International Comparison Project, is 'real' GDP, based on purchasing power. This attempts to gauge what countries – or individuals – could buy if they shopped in a single world supermarket.

'Real' GDP is better than the conventional measure in that it makes allowance for different cost structures. A medical consultation is a medical consultation anywhere in the world, but it is more expensive compared to a hamburger in the U.S.A. than it is compared to a chapatti in India. The same is true generally of services in comparison with manufactures.

'Real' GDP shares many problems with its conventional cousin: the difficulty of conversion into a standard measure (in this case 'international dollars'); weaknesses in accounting for goods and services not normally traded commercially; insensitivity to the effects of economic activity on the environment. But it does put right some of the distortions engendered by that cousin, and it shows, in broad outline, that world purchasing power, although still shockingly concentrated in a few states, is more evenly spread amongst them than world income.

32 FIRST CHARGE

Saddam Hussein is building the world's biggest mosque in Baghdad, capital city of war-ravaged, starving Iraq. Pakistan, one of the poorest of states, is spending untold billions on nuclear weapons.

Governments worldwide, be they autocratic or democratic, spend vast sums on objects or projects of their desire, not of their citizens' choice. They spend money in our name, largely unconstrained by our wishes. A nuclear power station here, a major road through unspoilt country there, an airport in the bush or a seaport on the rocks. Worldwide, one quarter of the U.S.$7 trillion they spend goes on paying their own salaries, and another tenth on keeping their military forces fed, housed and armed, compared with just over a tenth on health or education.

33 FOOD POWER

National self-sufficiency in food production is not a sign of an adequate diet or even of freedom from hunger, but it is a pre-condition for them.

As Earl Butz, a combative U.S. Agricultural Secretary, once said in the 1980s, 'Food is a weapon. It is now one of the principal weapons in our negotiating kit'. As a weapon, it is used by big and small alike, at home (in Sudan, for example) and abroad (by the U.S. or the European Union).

It is a weapon few states are likely to lay down. Although average agricultural yields differ wildly from state to state – from 6,650 kilograms of cereal per hectare in the Netherlands to 380 in Angola in the late 1980s – which leaves plenty of room for improvement in the world as a whole. No conceivable growth in productivity, even coupled with an extension of land available for food crops, will make all states self-sufficient within their current borders.

34 INDUSTRIAL POWER

The latest statistics available have not caught up with the breakup of the USSR and of Yugoslavia, with German reunification or with the independence of Eritrea. They nonetheless reveal how relatively small an industrial power the former USSR, let alone Yugoslavia, was, and how sharply the industrial power of East Germany had been declining during the 1980s. German reunification itself demonstrated how hollow much industrial production in East Germany was, as state enterprises unable to compete effectively in world markets were either abandoned or modernized at enormous cost in new investment. In Russia, as in other successor states of the USSR, much of the industrial economy is now little more than a matter of stage management. And, meanwhile, industrial growth in some Asian states has been continuing apace.

35 PRODUCT POWER

In 1992, the value of exported goods worldwide totalled U.S. $3.73 trillion, almost double the 1982 total of U.S. $1.88 trillion.

Asian states have shown the most dramatic increases in exports, and their economies have thrived on a combination of relatively cheap labour and a high rate of investment in technology, skills and initiative. Hong Kong and Singapore have promoted a large trade in re-exports – goods which are imported and then, without being transformed, profitably exported again. In 1992, Hong Kong's trade in re-exports amounted to 75 percent of its total trade in exported goods.

The steep decline in the world market price of oil had a corresponding impact on those states for which oil is their paramount export. Bahrain, Ecuador, Gabon, Qatar, Trinidad and Tobago would have featured in a 1982 version of this cartogram but they do not feature ten years later. Bulgaria, which is among those states whose economies imploded with the collapse of the USSR, is absent; so, too, is Iraq, a state subject to international economic sanctions.

36 SERVICE POWER

In 1992, the value of worldwide exports in commercial services totalled U.S. $1 trillion, or well over double the total in 1982. Even so, the recorded statistics significantly understate their real value. Some transactions, such as the repatriated earnings of migrant workers, may be excluded as remittances; others defined as trade in goods rather than in services; and more importantly, still others, transmitted electronically, often go unrecorded, especially when they are transactions between parent companies and their foreign affiliates. It is a sector in which the sale of information, banking and other financial services, including insurance, is a particularly fast growing element of the world economy. This explains why the dominance of the U.S.A., with 16.2 percent of the world total, is even more marked than it is for the trade in goods, where it is 12 percent. The share of France, at 10.2 percent is also markedly larger than its share of the trade in goods, at 6.3 percent.

37 SCIENCE POWER

In the world of cyberspace, Internet and e-mail, not everyone would choose to measure science output by the number of articles published in learned journals, or gauge the influence of a scientific observation by the number of references to an article in subsequent journal articles. But the scientists themselves do so, and we have no option but to follow them.

The measure imparts a number of biases to the results: English-language publications score better than publications in other languages; Latin script is favoured over others; disciplines with a few major journals (physics and earth sciences, or engineering and technology, for example) fare better than those with a relatively dispersed literature (biology and mathematics).

The conclusion is nonetheless clear: whatever the measurement used, citations, articles or royalties and fees, the U.S.A. is the science superpower, Japan is a rapidly rising sun, and there is a small cluster of European contenders.

38 MILITARY POWER

The Cold War may be over but arms spending still gobbles up a tenth of all government expenditure worldwide. More than two-fifths of the total is spent by the U.S.A. Military spending is as shrouded in secrecy and arbitrary pricing as it always has been. China's declared military budget of U.S. $6 billion a year, for example, could be anything between U.S. $28.5 billion and U.S. $45 billion in reality, depending on the method of calculation.

39 UNDER ARMS

The end of the Cold War in 1989 inaugurated a wonderful period for war-mongers. Of the 136 states for which the comparison can be made, no fewer than 75, that is 55 percent, increased the size of their armed forces between 1985, when the Cold War was at its height, and 1994, the last year for which figures are available. Despite heavy reductions in some of the major military forces, in NATO and the countries of the former Warsaw Pact, the number of people under arms worldwide has increased by 36 percent, from 22 million to 30 million.

Included in the cartogram are all military servicemen and women on full-time duty; full-time active members of paramilitary forces whose training, organization, equipment and control suggest that they are used in support or in lieu of regular military forces; and active internal opposition forces that come into these categories. All Reserves are excluded, even in states such as Israel or Switzerland where they constitute a prime ingredient of military strategy. National guards are excluded as are civil defence forces, people's militias, secret services and all other covert and overt forces employed in safeguarding the security of the state. States whose armed forces amount to less than 0.01 percent of the world total are also excluded.

Some heavily militarized states make unusually heavy demands on their manpower by employing a large core of professionals in their conscripted armed forces.

40 STATE TERROR

States that resort to capital punishment or that promote, condone or simply do not expunge the use of assassination, 'disappearances', torture or other cruel, inhuman or degrading treatment are terrorist organizations – in the sense that they, often systematically, employ violence and the threat of violence to influence the behaviour of people unnamed or unknown.

State terror can be vicious, like Iran's or China's; it can be self-righteous like Israel's; or shamefaced like Australia's. States can be shedding the most gruesome aspects of their past as they are in Eastern Europe, Paraguay, and Mexico; or clinging to them as in Nigeria. Some use terror lavishly in the interests of its ruling groups, as in Iraq; others like Sweden, are minor players, which officially condemn it, occasionally condone it and seldom, if ever, prescribe it. There are some states where it is strictly rationed, as in the UK; and others where it is copiously available, as in India. And there are states in which the central authorities are blameless in intention but subverted by local government – Western Samoa or Bulgaria.

41 SMOKING GUNS

Many things make wars more likely and more destructive. Not least among them are the trade in arms and arms' accumulation. Although the Cold War has gone, hot wars proliferate. War itself is changing. In the early part of the 20th century, nine-tenths of war casualties wore uniform. This reflected the fact that most wars were between sovereign states, and were fought by trained men dedicated to the purpose. The prudent civilian could assume that he or she would be able to keep out of harm's way and survive the conflict.

This is no longer true. War is now primarily a domestic affair, waged between a state and its challengers. Civilian casualties outnumber military victims by nine to one. In most theatres of war, the prudent civilian wishing to survive would do well to join the armed forces.

The study of war cannot be an exact science. As the focus shifts from formal, structured, interstate conflicts waged by named agents with clear duties to record and report, to civil war, beginnings and ends become fuzzy. Casualty and refugee numbers become fuzzier. Judgement mingles with measurement, and definitional struggles colour reality. Is the person dying of cholera in a Rwandan refugee camp in Zaire a war casualty or a victim of disease? And what about the young woman scarred by multiple rape in Bosnia? Is she a casualty of war, or of something else?

42 REFUGEE MAKERS

Nobody knows how many people have been compelled to leave their homes for fear of persecution, injury or death, and would return if they could or dared. They must number hundreds of millions, huddled in the world's slums, swarming across frontiers and oceans, quarantined in special camps. Although they are fugitives from violence of one sort or another and from discrimination and deprivation, most of these people are not classed as refugees in the restricted sense employed by the agencies concerned with them.

To qualify as a refugee in that sense, a person must have crossed a recognized interstate boundary in his or her flight and also be in need of assistance and protection. Relatively few people satisfy both criteria.

Those that do qualify do not include the 'internally displaced': that is, refugees within their state's own frontiers. They do not include the vast mass in flight from destitution and debilitation; the growing army of 'asylum seekers' waiting for registration as 'real refugees'; the people in 'refugee-like circumstances' who are undocumented, unregistered or who, for some reason, fall outside the official protection mechanisms of host countries and international agencies; the forcibly relocated by government resettlement programmes; the small minority of 'real refugees who, welcome or not, miserable or desolate, have 'entered and resettled'; and people in a host of other categories and sub-categories.

The cartogram rests on a broader, though still restrictive, definition which includes as refugees 'asylum-seekers in need of assistance and/or protection' (that is, people 'unable or unwilling to repatriate due to fear of persecution and violence in their homelands or to be permanently settled in other countries') plus selected 'populations in refugee-like circumstances' plus the 'internally-displaced' and the 'forcibly located' (as given by the U.S. Committee for Refugees).

We have not included the vast array of fugitives who have stolen away rather than stampeded from trouble. Amongst them are the people who would not have left their homes in the first place and would almost certainly return if there were any hope there of elementary material security. These are the archetypal 'economic refugees': people forcibly returned in their thousands by the U.S. to Haiti, Cuba and Mexico, by the British (in Hong Kong) to Vietnam, by the Saudis to Ethiopia. They are the people forcibly blocked by Austria on its borders with the former Communist states of Eastern Europe; by Finland at its frontier with Russia; by Western Europe as a whole in its southern and eastern reaches; by Turkey on its border with Iraq.

We have also excluded the 'returnees': three-quarters of a million or so Russian 'Jews' to Israel; the one and a half million ethnic and not-so-ethnic Germans who trailed into the then Federal Republic in 1989 and 1990; the vast numbers who manage, against all odds, to get to their chosen destination without attracting attention – the multitudes of illegals in France, in Italy, in Spain, in the U.S.A. And then there are the officially 'entered and resettled'.

People who are displaced by accident or design are included only exceptionally. They are the victims of irrigation projects or other 'economic development' programmes, of industrial or 'natural' disasters, and more.

Refugees are created. In a world of states, it is usually the state that creates them, primarily through war and the collateral effects of war, but also through forms of discrimination which amount to systematic violence, or through bad husbandry. It is for this reason that the cartogram reverses the conventional presentation by focussing on the source states rather than on asylum states.

No states are rushing to offer havens even to 'real refugees'. As refugees multiply, they find it increasingly difficult to go forth legally, let alone with a semblance of dignity. Recipient states resort to subterfuges which would be laughable if their consequences were not so tragic.

Amnesty International reports from France that an Algerian policeman and his young wife, who fled after receiving death threats from the 'armed Islamic movement', were refused refugee status on grounds that 'manifestly, they have nothing to fear from the Algerian state'.

In Belgium, lawyers representing detained asylum seekers, having obtained a court order requiring the government to give them 48 hours' notice of the date and time of appeal hearings, found themselves stymied by the minister of the interior who announced that he would fight this 'absurd decision' all the way on appeal and, if necessary, would bring in legislation to overturn it.

Even in the Netherlands, usually more humane than most, 35 Zairean refugees were on the point of being sent back to Togo because, the Dutch foreign ministry stated formally, it had been assured by the representative of the UN High Commissioner for Refugees in Lome, the Togolese capital, that it was safe enough for them to return there. The UNHCR has no representative in Lome.

Airlines face fines if they bring into a country people who do not have the required visa and travel documents. Many bona fide refugees do not have travel documents, and many states have imposed visa requirements with the express purpose of keeping refugees out. The British Refugee Council records the case of the Bosnian refugee who was unable to join his wife and children in Switzerland because he had no passport. Instead he went to the UK under a scheme for wounded or disabled Bosnians, but after twelve months he had still not been given refugee status, and his wife and children had not been allowed to join him. This is only one of thousands of Catch 22s.

43 DEBT JEOPARDY

Few governments seriously attempt to balance their books except when foreign exchange and credit markets require them to do so. In general, they spend more than they have the ability or the audacity to extract in taxes. The cost of servicing accumulated debts is the first charge on a government's resources: the higher the burden of debt, the lower the proportion of revenue available for everything else.

Poor states face double jeopardy. In periods of easy credit – notably, when the international banks were awash with money deposited by the oil rich states – they were encouraged to borrow from abroad huge sums in relation to their own resources. That much of this money may have been misappropriated or wasted does not make the accumulated burden of servicing and repayment any less exacting. Even if some of this debt had been rescheduled or cancelled, what remains is an often crippling charge on foreign earnings.

The scale of official development assistance to poor states reveals how little the rich states are ready to provide in relation to their riches.

44 FUNNY MONEY

Inflation, like taxation, is as certain as death. Indeed, it is increasingly a furtive form of tax, by which the state commands goods, services or credit for which it later pays in depreciated currency. Inflation is socially more iniquitous than taxation because its prime victims tend to be the most vulnerable, such as the elderly who rely on their savings; workers without the power to ensure that their wages keep pace with the rise in prices; and all those whose poverty allows them no means of coping. It is only the few with the necessary skills and financial resources who can take action to neutralize the process or even profit from it. In extreme but not uncommon cases – notably now among the successor states of the former USSR – inflation becomes hyperinflation, when money loses its value so fast that multitudes are reduced to destitution.

45 GOD AND CAESAR

Even unimpeachably secular states, which deal evenhandedly with all orthodox faiths or ideologies, profess values which shape their activities – the rule of law, belief as a matter for individual choice, and the individual as the ultimate repository of sovereignty. Most states are less reserved. They range from the secular in theory but partisan in practice, through the committed but tolerant, to the fundamentalist.

Some of them, such as the ones that profess 'Marxism-Leninism', like China, or extreme, mystical nationalism, like Myanmar, uphold beliefs which are not those of their people. Others, which seem to uphold the prevailing popular beliefs, do so only in the most general sense. In practice the governing classes profess a distinct variant of the prevailing belief, as in large parts of Latin America or the Middle East. The diffusion of a handful of world cults has resulted in a doctrinal tapestry, each varied enough to cover the most incompatible social and political purposes.

State beliefs have not been unaffected by recent shifts in world power. 'Marxism-Leninism' as an official doctrine, with its prophets, holy texts, licensed interpreters, shrines, rituals and sacraments, is in steep decline. Mystical nationalism, with a similar spiritual armoury, is on the rise. At a popular level, volatile social change, experienced as personal insecurity, has resulted in an unprecedented and continuing surge of fundamentalism characterized by God-sanctioned militancy, a search for alternative religious and political institutions (which is putting the current incumbents under growing pressure), and the pretence of working within an unbroken tradition.

Claims that are made about the number of adherents to a belief should be taken with a pillar of salt. Confessions of faith are made in response to someone else's demands, and are, like taxes, paid willy-nilly. But unlike taxes payment is made in spiritual currency and cannot easily be gauged. That being the case, this atlas has reluctantly followed the religious encyclopaedists who, for obvious reasons, assign to everyone – babes in arms and dodderers in bed included – a coherent system of beliefs. The result is that numbers and, by implication, the level of general commitment to conventional faiths are grossly overstated.

46 PROGRAMMED VIEWS

Television, with its cultural and often subliminal impact, may well be the most potent form of mass communication. Though regulated to provide so-called balanced comment in some countries, in others it is a critical instrument in manipulating opinion. In poor countries, where relatively few sets may be privately available, community sets or those installed in cafés and bars, greatly augment the number of viewers.

Radio remains everywhere a significant medium, and not only where television sets are scarce. In the U.S.A., for instance, radio has become a far more important medium for the expression of – mainly right-wing – political views, through hosted talk shows or stations.

Newspaper circulation depends on more than the mere extent of basic literacy, and in some poor countries – where the cost of buying a newspaper is beyond the means of all but a very few – the circulation rate per 1,000 of the population is so low that it is statistically recorded as zero. Yet even there, newspapers may be influential in affecting opinion within the elite. Elsewhere, and particularly in the advanced economies with more or less democratic systems of government, the press is still the most potent medium for the direct transmission of political influence.

47 THE THREE MONKEYS

There is a world of difference between sifting through opinions and information to find what might be apt for given circumstances or to make them digestible, and censorship – the outlawing of views and news by people dedicated to the task. The one is inseparable from social life, the other a naked exercise of social and political power.

Formally, most states renounced censorship when they adhered to the UN's Universal Declaration of Human Rights which, in Article 19, holds that 'Everyone has the right to freedom of opinion and expression; this right includes freedom to hold opinions without interference and to seek, receive and impart information and ideas through any media and regardless of frontiers.'

Declarations, of course, are one thing, and practice is another. States employ an array of instruments to suppress unwelcome opinion and prevent its dissemination: killings, kidnapping, disappearances; arrests, detention and imprisonment; dismissals, threats and harassment; restrictions on movement and expulsion; control of news agencies, setting of guidelines, favouritism, leaks, control of access, training of media staff, surveillance of press clubs, herding and corralling of journalists; disinformation; legislation, licensing, bans, expulsions; ownership; bribes; allocation of scarce resources including newsprint, frequencies, typewriters, advertising.

Standard targets for censorship include editors, producers, journalists, printers, writers, academics, human rights activists, political dissidents, and artists of every variety. And among a state's many justifications for censorship are the protection of innocents from unfair criticism, from defamation and from the invasion of privacy; the protection of the state or national security; the control of sedition; the protection of public health, public morals, public order; the preservation of a linguistic, cultural or public regime; the correction of media bias; and, the enforcement of contractual obligations.

The inset map covers prisoners of conscience as defined by Amnesty International: people detained for their beliefs, colour, sex, ethnic origin, language or religion who have not used or advocated violence, as well as people considered by Amnesty as possible prisoners of conscience. It includes conscientious objectors to military service, and prisoners adopted by International PEN. The data and judgements on which the map is based relate to the early 1990s. The anomalies reflect the times, and biases of our sources.

48 TONGUE-TIED

The queen of imperial languages is English, the nearest the world has come to a common tongue. By the mid-1980s English was claimed as their home language by more than 300 million people. Another 300 million or so used English as a second language, and a further 100 million used it fluently as a foreign language. This is an increase of 40 percent since the 1950s. Other estimates put the total number of English speakers, including people with a lower level of fluency, at over one billion, that is, well over a quarter of the world's teenage and adult population.

English is used as an official or semi-official language in over 70 states and has a prominent place in another 20. It is the premier language of books, newspapers, aviation and air traffic control, international business, academic conferences, science, technology, medicine, diplomacy, war, sports, international competitions, pop music and advertising. Over two-thirds of the world's scientists write in English. Three-quarters of the world's mail is written in English. Of all the information stored in electronic retrieval systems, 80 percent is stored in English. English-language radio broadcasts transmitted abroad are received by over 150 million people in 120 states. More than 50 million children study English as an additional language at primary school; and over 80 million study it at secondary level. (These figures exclude China where over 100 million people are thought to have tuned into the BBC Television English series 'Follow Me' at some time.) In any one year the British Council helps a quarter of a million foreign students learn English in various parts of the world. In the U.S.A., nearly half a million foreign students of English are registered every year. This dominance of English, and the other imperial languages, effectively squeezes out indigenous languages, as is shown on Map 49 Pressures.

The main map greatly exaggerates the linguistic power of the world's population. Language can be a barrier as well as a bridge. Although there are many people for whom the dominant language is not a mother tongue (or home language) yet who can use it fluently and so participate fully in public affairs, there are very many more who claim the ruling language as theirs and yet, by reason of poor education, narrow experience or inadequate nurturing, use it so uncomfortably or in such a distinctive form as to be excluded from effective participation.

In the case of some ex-colonial states it is not always easy to know the real status of an indigenous second official language. Whether its speakers have access to 'rule' depends partly on the extent to which the language is taught at school, on the prevalence of literacy, or on some such shoehorn factor. In all cases the indigenous language has been given the benefit of the doubt.

49 PRESSURES

Some states use the law to secure conformity with proclaimed norms of sexual expression. Others rely on social pressure and prevailing attitudes. In many states, self-appointed guardians exert their own pressures through more violent methods.

We have classified states according to broad categories. 'Not forbidden by law' means only that there are no discriminatory provisions against homosexuality. It does not imply an absence of pressures through social disapproval. 'Legal with some discriminatory provisions' encompasses lawfulness but with such restrictions as a higher age of consent for homosexual than for heterosexual practices; the banning of homosexuality in a particular sector, usually the armed forces; and/or the specific banning of anal intercourse between men. 'Lawful but repressed' is a category applied to states where the law is silent on homosexuality, but provisions against 'indecency' or other formulations are effectively directed against homosexuals; where the harassment of homosexuals by the police or vigilantes is evidently sanctioned by the authorities; and/or the social climate is extremely hostile. 'Unlawful but tolerated' means that the authorities seldom prosecute and that society is largely tolerant (though in some such cases, the rise of fundamentalism is beginning to threaten the traditional attitude). The two remaining categories need no elaboration.

Less proscriptive but even more potent is the pressure exercised by the dominant culture on minority ones, and especially on the languages involved. Minority languages are everywhere under threat, not least from the few languages spoken internationally (see Map 48 Tongue-tied). The American Association for the Advancement of Science conference at Atlanta in February 1995 was informed that only ten percent of the world's 6,000 languages are likely to survive the next 100 years. Already between 20 percent and 50 percent are no longer being learned by children and will die with their surviving speakers.

50 ADDITION AND DIVISION

And still they come. Since the last edition of this atlas was being prepared, in 1990, 23 new states have raised their flags and posted ambassadors abroad. Most of them joined the United Nations.

The nation state is one in which, ideally, the political unit coincides with a popular sense of common identity. It goes no farther back than the 18th century. Of the states that exist today, only 14 have had an uninterrupted existence for more than 200 years. From 660 BC, when Japan came into being as a recognizable entity, to 1776, with the constitution of the U.S.A., these pre-19th century states came into being on average once every 175 years. Then the pace of state formation accelerated: to an average of one state every four years in the 19th century (28 in all), one in every 18 months in the first half of the 20th century (37 in all), one in every five months in the latter half of the 20th century, up to 1993 (105 in all).

It is not hard to understand why the novelty of the nation state was taken up so widely and with such profound effect. The emerging market system needed the state to underpin it. So rest assured that San Marino (population 24,091, area 60 kms^2) joined the UN on 2 March 1992, as did Liechtenstein (population 30,281, area 160 kms^2) on 18 September 1990, alongside the successor states to the USSR and Yugoslavia. In March 1995, there were 184 internationally recognized states (not including those recognized by only one other state, such as the Turkish Republic of Northern Cyprus).

Some of the new states are reincarnations of older ones: many of the 1990s crop are successor states to the former USSR and Yugoslavia. Some are waiting in the wings: in Israel or Morocco, for example, where they are restrained by forced possession, or in Sudan, Somalia or Liberia, where they are hatching in conflict with each other. For others – Canada, Spain, India, Indonesia perhaps – the bonds that hold the current states together are fraying. True, there are states and territories on a different trajectory: Hong Kong and Taiwan moving towards China, the Koreas towards unification. But appearances may well prove deceptive. Few of them enjoy the favourable circumstances of Germany, which was re-united in October 1990.

Ahooja-Patel, Krishna 'Gender Distance among Countries' *Economic and Political Weekly*. Bombay. February 1993.

Amnesty International (AI). *Conscientious Objection to Military Service*. London: AI, January 1991.

Amnesty International. *Report*. London: AI, annual.

Barrett, David B. ed. *World Christian Encyclopedia: A Comparative Study of Churches and Religion in the Modern World, AD 1900-2000*. Nairobi, Oxford and New York: Oxford University Press, 1982.

Belfield, Richard, Christopher Hird and Sharon Kelly. *Murdoch: The Great Escape*. London: Warner Books, 1994.

Borders and Territorial Disputes. 3rd ed. Harlow, Essex: Longman, 1992.

Central Intelligence Agency (CIA). *World Factbook 1993* and *1994*. Washington, D.C.: CIA, 1993 and 1994.

Europa Regional Yearbooks. London: Europa Publications, various dates.

Europa World Yearbook 1993. London: Europa Publications, 1994.

General Agreement on Trade and Tariffs (GATT). *International Trade Statistics 1993*. Geneva: GATT, 1994.

Greenpeace International. *Climate Time Bomb*. Amsterdam: Greenpeace, 1994.

Grimes, Barbara F. ed. *Ethnologue: Languages of the World*. 11th ed. Dallas, Texas: Summer Institute of Linguistics, 1988.

Gunnemark, Erik V. *Countries, Peoples and Their Languages*. Gothenburg: The Geolinguistic Handbook, n.d. (early 1990s).

Hungarian Academy of Sciences. *Scientometrics*. Budapest: Akademiai Kiado; and Amsterdam, Oxford, New York and Tokyo: Elsevier Science BV, serial 1994.

Index on Censorship. London: (monthly).

International Boundaries Research Unit (IBRU). *Boundaries and Security Bulletin*. Durham: University of Durham, Department of Geography, serial.

International Confederation of Free Trade Unions (ICFTU). *Free Labour World*. Brussels: serial.

International Institute for Strategic Studies (IISS). *The Military Balance 1993-94*. and *1994-95*. London: Brassey's, 1993 and 1994.

International Labour Office (ILO). *Yearbook of Labour Statistics*. Geneva: ILO, annual.

International Monetary Fund (IMF). *Economic Reviews* for Kirgistan, Tajikistan, Uzbekistan, 1992; Armenia, Azerbaijan, Georgia, Kazakhstan, Turkmenistan, Ukraine, 1993. Washington, D.C.: IMF, relevant dates.

International Monetary Fund (IMF). *Government Finance Statistics Yearbook 1993*. Washington, D.C.: IMF, 1993.

International Monetary Fund (IMF). *World Economic Outlook*. Washington, D.C.: IMF, October 1993.

International PEN. *Writers in Prison Committee, Case List, January 1994*. London: International PEN, 1994.

International Road Federation. *World Road Statistics 1953* and *1988-92*. Geneva: International Road Federation, 1953.

Keesing's Record of World Events. Harlow: Essex: Longman, serial.

Kurian, George Thomas. *Encyclopedia of the Third World*. New York and Oxford: Facts on File, 1992.

McEvedy, Colin and Richard Jones. *Atlas of World Population History*. London: Allen Lane, 1978.

Marty, Martin E. and R. Scott Appleby. eds. *Fundamentalisms and the State: Remaking Politics, Economics and Militance*. Chicago: Chicago University Press, 1993.

Marty, Martin E. and R. Scott Appleby. eds. *Fundamentalisms Observed*. Chicago: Chicago University Press, 1993.

Minority Rights Group. *World Directory of Minorities*. Harlow, Essex: Longman, 1989.

Motor Industry of Great Britain. *World Automotive Statistics 1993*. London: Society of Motor Manufacturers and Traders, 1993.

National Science Foundation (NSF). *Science and Engineering Indicators 1993*. Washington, D.C.: USGPO, 1993.

News Corporation. Annual Report, 1993.

Nicolaides-Bouman, A.N.S. and Nicholas Wald, with Barbara Forey and Peter Lee. eds. *International Smoking Statistics*. London: Institute of Preventative Medicine; and Oxford, New York and Tokyo: Oxford University Press, 1993.

O'Brien, Joanne and Martin Palmer. *The State of Religion Atlas*. London and New York: Simon and Schuster, 1993.

Parekh, Bhiku. 'Some Reflections on the Hindu Diaspora'. *New Community*. Coventry, University of Warwick, Centre for Research in Ethnic Relations. July 1994.

Population Action International. *Expanding Access to Safe Abortion: Key Policy Issues*. Washington, D.C.: PAI, September 1993.

Population Action International. *World Access to Birth Control*. Washington, D.C.: PAI, September 1992.

Population Concern. *1993 World Population Data Sheet*. London: Population Concern, 1994.

Population Reference Bureau. *1993 World Population Data Sheet*. Washington D.C.: Population Concern, 1993.

Privatisation International. *Privatisation Yearbook*. London: Privatisation International, annual.

Republic of China (Taiwan). *Statistical Yearbook 1993*. (Taipei) Directorate-General of Budget, Accounts and Statistics, 1993.

Russia and Eurasia: Facts and Figures. 18. Florida, Academic International Press, 1993.

Segal, Ronald. *The Black Diaspora*. London: Faber & Faber; New York: Farrar, Straus & Giroux, 1995.

Shawcross, William. *Rupert Murdoch*. London: Chatto & Windus, 1992.

Sivard, Ruth Leger. *World Military and Social Expenditures 1993*. Washington, D.C.: World Priorities, 1993.

Smith, Dan. 'Dynamics of Contemporary Conflict: Consequences for Development Strategies' in Nina Greager and Dan Smith. eds. *Environment, Poverty, Conflict*. Oslo: International Peace Research Institute (PRIO), 1994.

Staple, Gregory C. ed. *Telegraphy 1993: Global Telecommunications Traffic Statistics and Commentary*. Washington, D.C.: Telegeography Inc., 1993.

Statesman's Year Book 1993-94. London: Macmillan, 1993.

Stockholm International Peace Research Institute (SIPRI). *SIPRI Yearbook 1993: World Armaments and Disarmament*. Oxford and New York: Oxford University Press, 1993.

Tielman, Rob and Hans Hammelburg. 'World Survey on the Social and Legal Position of Gays and Lesbians' in Aart Hendriks, Rob Tielman and Evert van der Veen. eds. *The Third Pink Book*. Buffalo, New York: Prometheus Books, 1993.

Union Postale Universelle. *Statistique des services postaux 1992*. Berne: Bureau International de l'Union Postale Universelle, 1993.

United Kingdom 1991 Census of Population. *Research and Statistics*. London: Department of the Home Office, 1991.

United Nations. *Energy Statistics Yearbook 1991 and 1992*. New York: UN, 1991 and 1992.

United Nations. *Monthly Bulletin of Statistics*. March 1994. New York: UN, 1994.

United Nations. *World Investment Report 1993: Transnational Corporations and Integrated International Production*. New York: UN, 1993.

United Nations Children's Fund (UNICEF). *The Progress of Nations 1994*. New York: UNICEF, 1994.

United Nations Development Programme (UNDP). *Human Development Report 1994*. New York and Oxford: Oxford University Press, 1994.

United Nations Educational, Scientific and Cultural Organization (UNESCO). *Statistical Yearbook 1993*. Paris: UNESCO, 1993.

United Nations Environment Programme (UNEP). *Environmental Data Report 1993-94*. Oxford: Blackwell, 1993.

United Nations Industrial Development Organization (UNIDO). *Industry and Development Global Report 1993-94*. Vienna: UNIDO, 1993.

United States Arms Control and Disarmament Agency (USACDA). *World Military Expenditures and Arms Transfers 1991-92*. Washington, D.C.: USCR, annual.

United States Committee for Refugees (USCR). *World Refugee Survey*. Washington, D.C.: USCR, annual.

United States Department of Agriculture (USDA). *Tobacco: World Markets and Trade*. Washington, D.C.: USDA, 1994.

United States House of Representatives Committee on Energy and Commerce, Subcommittee on Telecommunications and Finance. Transcript of Hearing on Derivatives, 19 May 1994.

United States State Department. *Country Reports on Human Rights Practices for 1993*. Washington, D.C.: USGPO, 1994.

Universal Postal Union (UPU). *Postal Statistics 1992*. Berne: UPU, 1994.

Wallensteen, Peter and Karin Axell. 'Conflict and Resolution and the End of the Cold War'. *Journal of Peace Research*. 31(3) August 1994. London, Newbury Park and New Delhi: SAGE Publications (for PRIO).

World Bank. *Atlas 1994*. Washington, D.C.: World Bank, 1994.

World Bank. *The Environmental Data Book*. Washington, D.C.: World Bank, 1993.

World Bank. *Social Indicators of Development 1994*. Baltimore and London: Johns Hopkins University Press, 1994.

World Bank. *World Development Report 1993*. Washington, D.C.: World Bank, 1994.

World Bank. *World Tables 1992*. Baltimore and London: Johns Hopkins University Press, 1992.

World Directory of Diplomatic Representation. London: Europa Publications, 1992.

World Energy Council. *Report 1993: International Energy Data*. London: World Energy Council, 1993.

World Resources Institute. *World Resources 1991-92* and *1994-95*. New York and Oxford: Oxford University Press, 1992 and 1994.

World Tourist Office (WTO). *Yearbook of Tourism Statistics 1993*. Madrid: WTO, 1993.

In addition, we have been greatly helped by personal communications, not all of them cited in the acknowledgements on page 11; by press reports, principally in *The Economist* (in particular, of 16 March 1994), *Financial Times* (in particular, of 20 October 1993, 2 December 1993, 7 January 1994), *The Guardian* (in particular, of 25 June 1994, 1 August 1994, 20 February 1995), and *The Independent* (in particular, of 15 May 1994); and by press releases and database searches, notably from GATT, WHO and WTO.